"To effectively mitigate against climate change, the U.S. needs coherent policies that support the delivery of clean energy to the market. Filled with lucid and poignant examples, this book thoughtfully shows how our often-conflicted regulatory environment impedes the efforts of clean energy entrepreneurs to implement their ideas."

—Eric Coleman, *The Florida State University*

Green vs. Green

Renewable and carbon-neutral energy have been promoted as the future of energy production in the United States. Non-traditional energy sources show promise as alternatives to fossil fuels and may provide a sustainable source of energy in increasingly uncertain energy markets. However, these new sources of energy face their own set of political, administrative, and legal challenges. *Green vs. Green* explores how mixed land ownership and existing law and regulation present serious challenges to the development of alternative energy sources in the United States.

Analytically examining and comparing five green energy sectors; wind, solar, geothermal, biofuel and hydro power, Ryan M. Yonk, Randy T. Simmons, and Brian Steed argue that discussing alternative energy without understanding these pitfalls creates unrealistic expectations regarding the ability to substitute "green" energy for traditional sources. The micro-goals of protecting individual areas, species, small-scale ecosystems, and other local environmental aims often limits ability to achieve macro-goals like preventing global climate change or transitioning to large-scale green energy production. Statutes and regulations designed to protect environmental and cultural integrity from degradation directly conflict with other stated environmental ends. Although there is substantial interest in adding clean energy to the grid, it appears that localized environmental interests interfere with broader environmental policy goals and the application of existing environmental laws and regulations may push us closer to gridlock.

Green vs. Green provides a fascinating look into how existing environmental law created or will create substantial regulatory hurdles for future energy generations.

Ryan M. Yonk is Assistant Professor of Political Science at the Department of Political Science and Criminal Justice, Southern Utah University and Research Director of the Center for Public Lands and Rural Economics, Utah State University.

Randy T. Simmons is Professor of Economics and Director of the Institute of Political Economy at Utah State University's Jon M. Huntsman School of Business, Senior Fellow at the Independent Institute, and former Mayor of Providence, Utah. He is a member of the Board of Directors of the Utah League of Cities and Towns and a Member of the Utah Governor's Privatization Commission.

Brian C. Steed is Instructor of Economics at the Huntsman School of Business at Utah State University, where he teaches in the Department of Economics. As an academic, Dr. Steed has worked on projects sponsored by various governmental and non-governmental agencies. His research examines the intersection of law, economics and policy, with a particular focus on land, resources, energy, the environment, and international development.

Routledge Research in Environmental Policy and Politics

1 **Green vs. Green**
The Political, Legal, and
Administrative Pitfalls Facing
Green Energy Production
*Ryan M. Yonk, Randy T. Simmons,
and Brian C. Steed*

Green vs. Green

The Political, Legal, and Administrative Pitfalls Facing Green Energy Production

Ryan M. Yonk, Randy T. Simmons, and Brian C. Steed

Routledge
Taylor & Francis Group

LONDON AND NEW YORK

First published 2013
by Routledge
711 Third Avenue, New York, NY 10017

Simultaneously published in the UK
by Routledge
2 Park Square, Milton Park, Abingdon, Oxfordshire OX14 4RN

First issued in paperback 2014

*Routledge is an imprint of the Taylor & Francis Group,
an informa business*

Library of Congress Cataloging-in-Publication Data
Yonk, Ryan M.
  Green vs. green : the political, legal, and administrative pitfalls facing green
energy production / Ryan M. Yonk, Randy T Simmons, Brian C. Steed.
    p. cm. — (Routledge research in environmental policy and politics ; 1)
  1. Renewable energy sources—United States.   2. Energy policy—
Environmental aspects—United States.   I. Simmons, Randy T.   II. Steed,
Brian C.   III. Title.
  HD9502.U52Y66 2012
  333.79'4—dc23
  2012010569

ISBN 978-0-415-53127-6 (hbk)
ISBN 978-1-138-88654-4 (pbk)
ISBN 978-0-203-11613-5 (ebk)

Typeset in Sabon
by IBT Global.

Contents

List of Figures and Tables

FIGURES

TABLES

Acknowledgments

Similar to our other collaborative projects, this book would not have been possible without a dedicated research team and supportive staff. As the idea for this book began to grow out of our previous research ventures and our trip around Utah's renewable energy plants, our research team continued to support our massive undertaking. Hours upon hours were spent tediously reviewing the United States' Code, Code of Federal Regulations, Legal Cases, Environmental Impact Statements, and many more monotonous documents. In addition, we thank the staff at both Southern Utah University and Utah State University for supporting the idea and doing their upmost to ensure we had all the tools necessary.

To our researchers at Southern Utah University, when deadlines were close, the hours were late, and the work was bland, you always stepped up and worked. We want to recognize the great work done by: Traves Bills aka Bubba, Rhett Busk, Matthew Coates, Heidi Eysser, Nick Hilton, Amy McIff, and Neal Mason. We appreciate your willingness to work with us, meet deadlines, and your mastery of the material.

Our research team at Utah State University deserves special attention for their readiness to being managed from afar when necessary. We thank you for the late nights and hours spent in front of a computer, reading and editing. Thank you for completing last-minute projects, answering our phone calls and emails, and putting in the time to make this book possible. This research team included: Joshua Blotter, Richard Criddle, Joshua DeFriez, Danika Foley, Luci Griffiths, Dan Groberg, Justine Larsen, R. Christopher Martin, and Sarah Reale.

Of course we need to thank Kayla Harris. Without Kayla this book would not exist. Kayla was willing to drop everything and devote herself to making sure that this book was finished. Not only did Kayla spend hours editing, but also she was willing to pick up any slack, from both team members and especially from us. Kayla put her life on hold in order to get this book done, and we are indebted to her for her hard work, dedication, and patience.

The book benefited greatly from the work of Kristen Dawson. We would like to thank her for her detailed storytelling, her willingness to edit or write at a moment's notice, and her management of the researchers at Utah State University. Kristen has put many hours into the perfection of all the small details in this book. She has included her shrewd insights, dynamic voice, and personality.

Supporting all of us are some fantastic staff, at both SUU and USU. At SUU we would particularly like to thank Provost Bradley Cook, department heads Michael Stathis and David Admire, and, of course, the dean, James McDonald, for their willingness to provide material support to this project.

Thus to all—we thank you and express our appreciation for your hard work and dedication; without you this project would not have been possible.

1 Introduction

During the 2011 State of the Union Address, President Obama made the aggressive challenge to obtain 80% of electricity development from "clean" sources by 2035 (Obama, 2011). Although the President was not fully clear on what he meant by clean energy, traditional sources of clean energy include alternative sources such as solar, wind, geothermal, and hydropower. Beyond these traditional alternative sources of energy development are other sources such as biofuel and oil shale. We include a discussion of the regulatory impediments to these alternative energy industries to illustrate the conflict in these energy sectors. The President has justified the need for such a massive policy shift based on the anticipated consequences of continued reliance on fossil fuels including geopolitical unrest in the areas of production of energy and the potential harm stemming from global climate change.

Reaching the President's goal will require substantial investment and development in clean energy infrastructure. In 2010, alternative energy made up just 11% of America's energy consumption (Romm, 2011). The World Bank has estimated that in the next forty years the amount of electricity demanded will increase by five million megawatts (MW) (Painuly, 2001). Additionally, the International Energy Agency predicts that if policies do not change, 90% of all energy demanded in 2020 will come from fossil fuels.

Historically, many have viewed technological shortcomings of existing clean energy generation processes as the primary barriers to alternative energy development. Although alternative energy production limits the release of greenhouse gases and other pollutants, it is well documented that alternative energy technology is far from achieving the efficiency rates necessary to supplant existing 'dirty' energies. Recently, however, alternative energies have seen substantial leaps in technology. Wind power capacity, for example, has increased from two to twenty gigawatts (GW), sound levels have halved, and capital costs are expected to decrease by 50–75% in the next twenty years (Bauen, Gross, & Leach, 2003). Additionally, economies of scale are moving production out of laboratories and into the market.

Solving technological problems might greatly increase alternative energy development. There are however, other, real barriers to clean energy development. After extensive review of current environmental policies, including those promoting alternative energy developments, we find that environmental laws and regulations hamper the development of alternative energy. The micro-goal of protecting individual areas, species, small-scale ecosystems, and other local environmental aims, often limits ability to achieve macro-goals like de-carbonizing the energy sector of the economy or transitioning to large-scale alternative energy production. Statutes and regulations designed to protect environmental and cultural integrity from degradation directly conflict with other environmental goals.

Like the technological challenges that face the clean energy sector, the economics of energy production from alternative sources remains a source of problems for energy producers. The reality is that most alternative energy production occurs at a significantly higher cost than traditional energy generation. Despite these higher costs, there is some evidence that the overall trend in many alternative energy production sectors is towards reduced costs and closer alignment with the market prices of traditional energy. Despite these trends, until alternative energy approaches direct parity with traditional energy costs, substantial development of alternative energy will remain limited. We do not address, however, the direct economics of alternative energy in the chapters that follow, but focus instead on the regulatory impediments that impact alternative energy development.

Our purpose in this book is to evaluate the political, legal, and regulatory environment in which large-scale clean energy development in the United States occurs. It is not to consider whether the claims made by proponents of particular energy sectors are correct or even plausible. In each chapter we outline the claims of the energy sectors proponents to illustrate the enthusiasm and aspirations of those proponents. These aspirations are then contrasted with the regulatory environment. We focus, in particular, on the political, legal, and regulatory environment for solar, wind, geothermal, biofuels, small to large hydro, and oil shale energy development. Our basic premise is that although some "green" conservation groups may advocate for developing alternative energy sources, when that development occurs near them, these group tend to criticize the action and lobby against the development. Thus, a green group versus green energy conflict ensues.

Although this green vs. green problem has received some attention in the popular press, it has been underexplored in both academic and policy settings. Existing literature on alternative energy falls into four broad categories: 1) technological developments; 2) comparative international analysis of climate policies; 3) examination of policies promoting reduction of greenhouse gases; and 4) economic analysis of general energy policies including alternative energy. Some attention has been paid to problems arising from public perceptions of clean energy siting and other NIMBY problems, but very little attention has been given to how existing regulatory regimes in the United States interact with the broad goals of developing alternative energy and de-carbonizing the economy.

To evaluate the potential effect of environmental regulation on alternative energy production we have conducted a systematic review of environmental statutes in the United States Code. Through this process we identified several hundred statutes with the potential to affect alternative energy. In making this determination, we considered possibilities of impacting siting, energy generation, and transmission. Examples of statutes likely to impact alternative energy include: the Endangered Species Act, the National Environmental Policy Act, the Wilderness Act, the National Historic Preservation Act, and the Indian Sacred Sites Act, just to name a few. In the Code of Federal Regulations (CFR), we identified thousands of regulations that govern the implementation of the statutes, and thereby potentially govern the development of alternative energy. We concluded that, in sum, environmental laws and regulations hamper the development of alternative energy. In addition, the micro-goal of protecting individual areas, species, small-scale ecosystems, and other local environmental aims often limits the ability to achieve macro-goals, like preventing global climate change or transitioning to large-scale clean energy production. Statutes and regulations designed to protect environmental and cultural integrity from degradation directly conflict with other environmental ends.

During the summer of 2011, we visited several alternative energy production sites to provide us with a more direct understanding of the industry from the ground. While touring a geothermal plant near Milford, Utah, we found an example of the direct impact these regulations can have on the provision of alternative energy. As part of the tour, the local manager of the geothermal plant took us to the top of the cooling tower, which is meant to reduce the temperature of the geothermal water before it is pumped back into the ground. As the local manager was explaining the mechanics of the cooling process, we looked out over the landscape and saw the plant's transmission line crossing the Utah desert. What was peculiar about this transmission line is the path of the line; it periodically made abrupt right then left turns, zigzagging across the desert as far as we could see.

Out of curiosity we asked the manager, "Why on earth does the transmission line zigzag as it leaves the plant?" The manager's response, "Are you familiar with the checkerboard pattern?" left us a bit perplexed until he explained further. The checkerboard pattern that exists across most of the west is the result of mixed estate landholdings, where survey sections (square parcels of 640 acres) were given to states primarily for education or to rail companies as incentives for the expansion of rail into the west during the nineteenth century. We pressed the manager to explain how this pattern of landownership explained the odd pattern of the transmission line. What was happening, the manager explained, was that the transmission line was carefully avoiding the federally owned portions of the checkerboard, favoring instead state or private lands.

After hearing this explanation, it was immediately clear to us that the geothermal company was studiously avoiding arduous federal regulatory processes by crossing only non-federal lands. When we pressed the manager as to why the company made this choice, he explained that it was far cheaper and much faster to incur the costs of a longer transmission line that zigzagged rather than comply with the full bevy of federal regulations required for crossing federal land. In this case, the energy developer was fortunate that it was possible to route the transmission lines solely on non-federal property. Most developments, at least in the West, are not so fortunate and face the reality of full compliance in order to generate any energy.

In the following chapters, we cover the most prominent alternative forms of energy. Each chapter follows a basic outline. We attempt to educate the reader about the history behind each energy type. Then, we review the major policies and legislation that have had a great effect on its development. Discussed as well are the advantages, disadvantages, and issues that still require attention. Each chapter includes a detailed, narrative case study of the interaction between alternative energy aspirations and green regulations.

In addition to federal laws and regulations, energy manufacturers have to deal with state regulations, which may be more stringent than the federal rules. California's endangered species regulations, for example, are more restrictive than the federal regulations. In addition to state rules, there are also local environmental laws. These can range from city or county regulations to homeowner associations (HOAs), which may restrict whether a homeowner can put solar panels on their roof or not. Developing alternative energy is a multifaceted puzzle, layered with complex and seemingly contradictory rules. Unfortunately, this complexity is something that is glossed over in grand public speeches over the necessity to reach alternative energy goals.

We have organized this book to first provide an introduction to the approach and theory we use to evaluate each of the potential alternative energy sectors. We next discuss the history, purposes, and impacts federal regulations have on developing each sector. After laying this foundation we turn to a detailed evaluation of each sector using an illustrative case study.

UNDERSTANDING GREEN VS. GREEN

In the second chapter we explain the green vs. green problem and introduce the Institutional Analysis and Development framework as a tool for reviewing how different actors and their environments can affect the decision-making process. We use Hardin's "Tragedy of the Commons" as a way of thinking about common-pool resources. Additionally, we outline how federalism, or the distribution of power between federal, state, and local governments, affects decisions.

REGULATIONS

In chapter three we briefly cover the major environmental regulations that have affected our case studies. Please keep in mind that the regulations we review are only a part of the legislation that would affect a potential alternative energy site. Each site, plan, and attempt to produce energy is likely to face idiosyncratic regulations that increase the regulatory complexity for the producers. Our approach in chapter three begins with the history of the legislation and then explores major litigation or consequences of the legislation. Through this chapter we demonstrate the proliferation of this legislation over time in order to provide an introductory review of several case studies that demonstrate the far-reaching effects of regulatory expansion.

Due to the scope and magnitude of the regulatory regime, it is impossible to provide a comprehensive summary of the impact of every regulation on the variety of alternative energy production sectors.[1] To explore the impacts of this regulatory environment, we develop illustrative case studies across six different sources of alternative energy production sectors. These sectors are: wind, solar, geothermal, biofuels, hydroelectricity, and oil shale production.[2]

In each case we provide a factual account of a project that has energy generation potential, but which has been hampered by existing environmental regulations. In selecting these cases we based our selection on four criteria. 1) The existence of energy generation potential at the selected site; 2) An actual proposal for energy generation at the proposed site; 3) That the site is representative of other power generation sites in the sector; and 4) That a detailed discussion of the proposal and site was available. Our fourth criterion was included to limit the possibility of selecting completely idiosyncratic cases.

ALTERNATIVE ENERGY SECTORS

Wind Power

Proponents of wind power see it as an essential part of a sustainable energy portfolio. Other alternatives to wind are technologically immature, only function in specific environments and therefore have limited development potential, are not cost competitive, or (in the case of hydroelectric power) the majority of potential sites have already been exploited (Hydroelectric power and water use, n.d.). Wind energy is "inexhaustible and infinitely renewable," does not use water during operation like other forms of energy, and public perceptions of wind energy are generally favorable (Wind Energy Benefits, Wind Powering America Fact Sheet Series, 2005, p.1–2).

Wind turbine generators (WTGs) can be located on land or offshore. Ideal sites are consistently windy, as intermittent power generation is one of wind power's shortcomings. Some optimistic analysts suggest that wind power has a major place in the future of energy production, accounting for as much as 20% of global consumption within two decades (DeCarolis & Keith, 2006). The United States has a staggering estimated 900,000 MW of offshore wind generation potential, particularly off the shallow waters of the Eastern seaboard, although there are currently no offshore wind projects online or under construction (Seelye, 2010, p. 1; Clarke et al., 2009). Despite technological improvements, significant government subsidies continue to be required for wind power to be economically competitive. Proponents believe, however, that it will soon be competitive with fossil fuels (Kennedy, 2005, p.1). Both on- and offshore potential wind farm sites are situated in complex regulatory and political environments, however, and often technologically and potentially economically viable sites are politically infeasible.

Case—Cape Wind

Since 2001, Cape Wind LLC has been attempting to site one hundred and thirty wind turbines off the coast of Cape Cod. One of the founders Jim Gordon grew up near this area and thought he understood the local culture. However, his plan to fill Nantucket Sound with wind turbines with an anticipated generating capacity of 454 MW, which in turn would power 160,000 homes surrounding the bay, was met with strong opposition. Local citizens protested and formed oppositions groups, while national leaders with land in the area also publicized their opposition. Aesthetic reasons motivated opponents to the Cape Wind installation, as the developers sited the project within view of several historical sites, affluent areas, and a landscape considered sacred to some Native American rituals, burial grounds, and relics (Clarke et al., 2009).

Like other energy production sites, offshore wind installations are subject to multiple levels of jurisdiction because of their location. Federal policies enter the equation because turbines are more than three miles from shore, while state and local policies govern the near-shore infrastructure (e.g. roads, power cables, etc.) (Clarke et al., 2009). These political chokepoints, most notably in the form of regulations and legal battles, have proven to be extremely difficult for the implementation of the Cape Wind project.

Solar Power

In 1954 Bell Laboratories patented the first high power silicon photovoltaic (PV) cell. This action caused the *New York Times* to project that one day solar power would harness the "limitless energy of the sun". But, growth in this sector has been slow; between 1990 and 2007 solar generation only grew by 3% on average (Schmalensee, 2010, p. 211). The International Energy Agency estimates that solar energy reached 1% of total energy production for the first time in 2009 for member countries of the Organization for Economic Co-operation and Development (OECD), a group of thirty countries located primarily in the northern hemisphere (Beereport,, Marmion, & Muller, 2011, p. 22). Unfortunately, certain barriers, such as transmission, technical and political issues have prevented solar energy from gaining a strong foothold in the energy sector.

Transmission will be an issue as long as population centers are located away from open, sun-intensive land. This problem is one of preference and cannot easily be changed. The second issue, technology, concerns the effectiveness of solar panels. Currently solar panels have an efficiency of only 6–15% and are projected to reach a mere 20% by 2020 (Zwaan, 2003, p. 3). The third issue, which will be explored more in-depth in the case study, is the political and regulatory climate that surrounds solar energy. Although policymakers have been discussing the benefits of alternative energy since the 1970s, little has been done to make the use of these fuels mainstream.

Case—Ivanpah Power

Luz International Ltd. began construction of the world's largest solar power plant in the 1980's. Construction was suddenly halted when the company came into financial difficulties and was forced to declare bankruptcy in 1991. Two decades later, under many of the same directors including the founder Arnold Goldman, BrightSource Energy is set to complete the Ivanpah Solar Electricity Generating Station (SEGS), in California. This plant is making headlines as it is "currently the world's largest concentrat[ed] solar power (CSP) plant under construction" (Ivanpah Solar Project Named CSP Project of the year, n.d.).

In order to produce solar power from its Ivanpah plant, BrightSource had to overcome several obstacles, specifically obstacles surrounding impacts of the Endangered Species Act's protection of the desert tortoise. The proposed solar power plant threatened the tortoises' natural habitat in California's Mojave Desert. To prevent negative impacts on the tortoise, construction plans have been frozen until BrightSource Energy finds a solution that protects the tortoises and satisfies green advocacy groups.

Geothermal Power

Geothermal energy harnesses the heat of the earth to produce electricity. There are multiple methods currently employed to extract this heat, but the process is essentially the same in each method. The first step is directing hot water from the earth's crust, which is pressurized and pushed upward from the earth's intense heat. From this hot water, steam is utilized to power a generator and create electricity. Proponents of geothermal energy boast that the capacity factors of constantly running geothermal plants are comparable to the energy output levels of coal and nuclear facilities while using less water, land, and other resources (National Renewable Energy Laboratory, n.d.; How Geothermal Energy Works, 2009).

The projected future of geothermal energy is staggering. In a recent assessment, the U.S. Geological Survey (USGS) stated that there is the potential of 8,000–73,000 Mega Watts (MW) on both public and private lands in sections across thirteen western states using current methods of geothermal energy utilization. In addition, the statement from USGS study found that hot "dry rock" resources could provide another 345,100–727,900 MW of energy capacity (How Geothermal Energy Works, 2009; Williams, et al., 2008). Geothermal energy "could one day supply nearly all of today's U.S. electricity needs," according to the Union of Concerned Scientists (2009).

Yet, even proponents of geothermal energy admit that there must be a technological leap before it can be a viable large-scale alternative energy option. At this point, the conventional technology is only used in areas with high plate tectonic activity, which makes up around ten percent of the earth's land area (How Geothermal Energy Works, 2009). They would need to be able to drill deeper in "dry rock" areas via the Enhanced Geothermal System to utilize all types of land and heat exchange areas (How Geothermal Energy Works, 2009).

Case—Telephone Flats and Fourmile Hill

On the Modoc Plateau, located in the Northeast corner of California, sits the Medicine Lake Caldera. According to the Pit River Tribe, after the creator finished making the earth, he and his son came to Medicine Lake (the Caldera's namesake) to bathe themselves from their exertions. While there, his presence imbued the water with a certain healing power, making the lake a holy place. For centuries, tribes ranging from the Pacific Ocean to the Rocky Mountains have sent their Medicine Men there to train and commune with the Great Spirit. Today, it is a beautiful destination for vacationers. Combining a camping experience with all the fun of motor boating, it provides a varied and enjoyable wilderness experience (Stanford Law School, n.d.). Yet, the commercialization of the lake's region isn't limited to tourism.

In recent decades, geologists have discovered that more can be offered from Medicine Lake than its previous use of healing waters and historical significance. A geological study performed by Stanford University faculty, and sponsored by Calpine Corporation (a geothermal energy development firm), describes the potential of this region. The study found that "[Medicine Lake] is perhaps the most promising, currently undeveloped, electrical-grade geothermal resource in the contiguous United States" (Hulen et. al., 2000). The region has sufficient geothermal heat and water resources to provide substantial amounts of alternative energy. In fact, Calpine has submitted plans to develop two plants in the Caldera. Each would provide 49.9 megawatts of energy. Additionally, Calpine has stated that over the next forty-five years, in which they plan to develop geothermal energy sites in the area, they expect to see as much as 1,000 megawatts of energy provided from these sites (Pit River Tribe, 2011). In this chapter, we tell how the Pit River Tribe Coalition has thwarted Calpine's plans.

Hydroelectric Power

The generation of electric power through the use of large and small-scale hydroelectric systems harnessing the kinetic energy of rivers and dams has been a reliable and cost-effective form of alternative energy (Belmans, Dragu, & Sels, 2011, p. 2). Some large-scale hydroelectric projects have created environmental problems while providing power. Smaller hydroelectric projects are, however, generally viewed as an environmentally sound way of developing energy (Schwartz & Shahidehpour, 2011, p. 2). These projects also represent the greatest opportunity for new sites in the United States (Paish, 2002, p. 539). An economically feasible hydroelectric site with little significant environmental impact may not be developed due to regulatory costs and barriers. Only 3% of the United State's 80,000 dams currently generate power (Ray, 2010). This low number is due at least partially to stringent regulations (Morris, 2011).

In the development of hydroelectric power a clear distinction emerges between technologically and economically feasible sites. A waterfall in a high mountain pass may be a great candidate for hydroelectric power, but limited transition corridors, and long distances may make the site economically infeasible. Our analysis suggests that there is another consideration beyond technology and economics: the regulatory environment surrounding a site may make it *politically* infeasible even if it is a good candidate technologically and economically.

Case—Glen Canyon Dam

Proposals for damming Glen Canyon emerged as early as 1916 and the Colorado River Compact of 1922 effectively divided the river's waters evenly between the upper and lower basin states. Construction for a dam was authorized but a site was not determined until 1946. When the Bureau of Reclamation published a study that identified Glen Canyon as a desirable site with promising hydropower capabilities. This declaration spurred a wave of environmental backlash, particularly from the Sierra Club, and protests slowed construction of many proposed dams. Amid years of heated controversy, Glen Canyon Dam was completed in 1963.

Despite the completion of the dam, environmental pressure to remove the dam quickly emerged and in 1989, the Secretary of Interior requested an Environmental Impact Statement to further evaluate claims by environmental groups that the dam was causing substantial environmental degradation. The EIS resulted in a reduction of the dam's output, which pleased environmental groups but failed to pacify them. As a result of this failure to satisfy all of the demands placed by environmental groups substantial pressure to alter or remove Glen Canyon Dam remains a volatile issue.

Biofuel

Biodiesel is a new take on the petroleum diesel that was being produced by the very late nineteenth century. It is renewable, energy efficient, and it displaces petroleum derived diesel fuel. It can be used in most diesel equipment with no or only minor modifications and it is produced with clean resources. It also has the advantage of combusting at a much lower temperature than traditional diesel, which in turn reduces greenhouse gas emissions.

The annual production of biodiesel in the United States reached nearly 700 million gallons in 2008, up from only 25 million gallons in 2004 (Brooks, 2009). Biodiesel is gaining wide acceptance as a mixture with conventional diesel fuel, with amounts up to 20% effective in unaltered diesel engines. The largest disadvantage when looking at biodiesel is the cost both in the form of the fuel itself as well as the impact it has upon crop prices, which are used for fuel production rather than consumption. Both in terms of emissions and the effects of spills, biodiesel is safer than traditional diesel. The safety of biodiesel is without question substantially higher than its competitors.

The larger effects of using biodiesel and ethanol are beginning to be understood. As such, some of unintended consequences have started a debate about the costs and benefits of this energy source. Unfortunately because of this, there is no case study to present for this chapter. In place of a narrative we instead discuss at length many of the issues and the controversies surrounding biodiesel and ethanol.

Oil Shale

Oil Shale refers to a sedimentary rock formed millions of years ago from organic matter that was then subjected to intense heat and pressure. This process is similar to that which formed most other fossil fuels, but in the case of oil shale it occurred incompletely resulting in a precursor petroleum-like substance. Pockets of oil shale can be found around the world, but the largest known deposit can be found in the Green River deposit, located in eastern Utah, southwestern Wyoming, and western Colorado. Rough estimates have indicated that this deposit contains more potential oil than the entire Saudi Arabian deposit. These estimates tend to range from 1.2 to 1.8 trillion barrels, of which somewhat more than half is estimated to be recoverable with existing technology (About Oil Shale, n.d.).

Oil shale was formed at lower temperatures and pressures than other fossil fuels. Once it is extracted it then must undergo a series of processing steps to transform it into a state similar to regular oil. Oil shale isn't currently a commercially viable resource because, as of the writing of this book, it is more expensive to process and use than regular petroleum sources. Attempts at making oil shale into a viable energy source have failed numerous times, due largely to economic issues. Oil shale also tends to have a laundry list of environmental problems stemming from extraction and processing of the resource, as well as its use, which is essentially similar to oil. Despite these environmental issues, we classify oil shale as an alternative energy, because it is replacement for traditional petrochemicals, and also because the case study demonstrates federal government use of environmental regulations to control their land holdings. Oil shale is currently listed as an energy source with a promising future, but before that future can be realized, numerous technological advances must be made.

Case—Uintah Basin, Orion v. Salazar

This case study deals directly with the area surrounding the Green River deposit in Utah's Uintah basin. The Uintah basin was inhabited by a succession of pre-Columbian peoples, ending with the Ute bands, which were the inhabitants when the first European explorers (the Escalante-Dominguez expedition) made their way into the basin. Trappers followed in the years after, who were then followed by the Mormon pioneers who had little use for the basin until the discovery of its many minerals. Following the discovery of oil shale, amongst other forms of minerals, the region experienced a mining boom that led to the development of some of the resources. This development caused the federal government to withdraw land from the Uintah reservation, which had been created by President Abraham Lincoln in 1861. Since this time, mining of various resources, such as natural gas, has come to define many of the towns found throughout the region. Oil shale production hasn't had much of an effect to date, but the interest that is currently brewing over the resource has potential to transform the basin and its culture.

CONCLUSIONS AND IMPLICATIONS

The cases introduced above and detailed in the chapters provide some cause for concern in the development of alternative energy sources. In each case, existing environmental law created (or will create) substantial regulatory hurdles for energy generation. Such regulatory barriers may be particularly problematic for clean energy development. Unlike some dirtier counterparts like coal, clean energy requires specific physical characteristics. Wind power can only be located in areas where there is sufficient wind. Large-scale solar facilities can only be located where there is sufficient land, sun, and water. Geothermal facilities can only be sited in locales with the appropriate geology, and hydro can only be located in areas with the appropriate hydrologic and geographic conditions. Adding the regulatory overlay to these physical requirements will certainly restrict widespread clean energy production as regulations take lands with the appropriate physical conditions off the table.

There are numerous clean energy projects proceeding throughout the country, within the boundaries created by existing environmental laws. Some caution is warranted, however. In the majority of the successful ongoing projects, it appears that energy companies have picked the low hanging fruit in selecting current sites for clean energy development. Sites have been selected that will reduce the possibility of conflict with existing environmental regulations. As additional energy projects are promoted, these easily developable areas are becoming scarcer.

For example, in the programmatic EIS developed by the BLM for solar energy production, the Bureau conducted widespread surveys of areas that could be readily developed for clean energy. After that extensive search, the BLM selected twenty-four separate sites that it proposes as Solar Energy Zones—those lands that are most readily developable based on land characteristics, limited environmental controversies, and access to existing power lines.

This process of selection is remarkable for several reasons. First, of the vast areas of land managed by the BLM, only twenty-four sites were considered readily developable for solar energy production without substantial controversy. Second, even the twenty-four proposed Solar Energy Zones have already faced challenges.

Almost everyone agrees that cleaner energy is in the long-term interest of the nation (and world). The problem is that there is little agreement about how and where to put cleaner energy facilities. Localized environmental interests interfere with broader environmental policy goals. Although there is substantial interest in adding clean energy to the grid, it appears that that the application of existing environmental laws and regulations may push us closer to gridlock.

2 An Institutional Framework for Analyzing Conflicts between Green Goals and Green Regulations

INTRODUCTION

Our purpose in this book is to evaluate the political, legal, and regulatory environment facing large-scale alternative energy development in the United States. In particular, we focus on the political, legal, and regulatory environment for solar, wind, geothermal, biofuels and small to micro hydro energy development. There are many evaluations of the economics of alternative energy and we do not attempt to add to or evaluate those claims (see, for example, Fogarty, Tom & Robert Lamb, 2012). Economic claims aside, there are real barriers to entry for developing alternative energy. The most prominent of these barriers are the existing environmental laws and regulations. After extensive review of current environmental policies, including those promoting alternative energy development, we argue that in sum, environmental laws and regulations hamper the development of clean energy. In addition, the micro-goal of protecting individual areas, species, small-scale ecosystems and other local environmental aims often limits the ability to achieve macro-goals like preventing global climate change or transitioning to large-scale alternative energy production. Statutes and regulations designed to protect environmental and cultural integrity from degradation directly conflict with other stated environmental ends.

We analyze political choices using the tools and assumptions of economics, namely to analyze how individuals choose within the constraints of rules. We assume that individuals- citizens, legislators, bureaucrats, and interest group members have passions and interests they pursue through governmental processes, and that these processes are used to reach specific goals. Individuals who pursue these goals do so because they are investing their time in doing what they value the most. Conversely, they choose to not invest in actions that have little payoff for them. Of course, we do not always know what is best for us, but neither does anyone else. We always choose in varying states of ignorance with respect to what nature and other rational, self-interested people are contemplating. But, people have invented ways of reducing uncertainty. We call those inventions rules; some rules are cultural, others are political implementations of cultural rules, and others are simply political.

The social order that emerges from socio-political rules reduces uncertainty and allows people to advance their self-interests. Such rules structure behavior in the political game. Among those rules are constitutions that prescribe how subsequent policy choices and decision changes are to be made. Civic rights, property rights, contracts, etc., serve to better define who commands which resources and how they may be employed and transferred. The important point is that collective institutions are highly important matters with good and bad consequences for individual and joint welfare.

INSTITUTIONAL ANALYSIS AND POLICY DESIGN

Evaluating the political, legal, and regulatory environment for green energy requires choosing an analytical framework, a theory of policy development, implementation, and evaluation. There is, however, no overarching or unified theory of public policy. In fact, there are competing and even conflicting theories. The February, 2009 issue of *The Policy Studies Journal* (Vol. 37, issue 1), for example, contained what the editors called a "policy shootout" in which authors of ten papers competed to make the best pitch for their favored approaches to policy analysis. Many of the proffered approaches overlap and complement each other. Even so, it is quickly clear from reading the articles that the field is an eclectic one.

We chose to apply the Institutional Analysis and Development (IAD) framework to policy analysis and design developed by Elinor Ostrom and her colleagues at the Workshop in Political Theory and Policy at Indiana University (Ostrom, 2011, 7–27; Ostrom, 2007, 21–64). The IAD framework is a policy analysis tool that provides a means of organizing information and data across different policy levels. It emphasizes organizations and interactions among the various actors in the policy arena (Ostrom calls it an "action arena") that interact. The IAD emphasizes what we believe to be the most fundamental aspect of policy formation—the institutions within which policy processes take place. We use "institution" in the way used by Ostrom (1996) and Douglass North (1990). An institution, according to them, is "a widely understood rule, norm or strategy that creates incentives for behavior in repetitive situations" (Polski & Ostrom, 1999). Institutions can be laws, policies, customs or formal and informal procedures. They may be visible, as in the case of a particular law, or they may be invisible, existing in the minds of each member of a community. They coordinate actions because they create expectations about how others will act in a particular policy or collective choice arena.

Both Ostrom and North stress that institutions and organizations are different analytical concepts. Institutions are rules. Organizations are structures. Institutions may be the product of conscious human choices such as a law, regulation, or court decision, or they may be the product of the unconscious accretion of rules of thumb or experiences. Institutions that are the product of human actions but not of conscious design can be as formal as the common law and as informal as the rules determining whether offense or defense calls fouls in a pick-up basketball game. Whereas organizations are visible, institutions can be rather invisible, especially to outsiders. A city council is a visible organization, for example, yet the customary rights, sanctions, and norms shared in a community are often invisible in the sense that they are not written down anywhere. They develop through time and exist in the minds of the community members.

Social order emerges because socio-political institutions and organizations reduce uncertainty and allow people to advance their self-interests. They structure the behavior in the political game. Political organizations create policy arenas within which rules are decided. Among those rules are constitutions that prescribe how subsequent policy choices and decision changes are to be made. Civic rights, property rights, contracts, etc., serve to better define who commands which resources and how they may be employed and transferred (Kiser & Ostrom, 1982, p. 208). Kiser and Ostrom identified three levels of political institutions: procedural, collective choice and constitutional. The procedural level consists of rules that direct how individuals act. The collective action level is the set of rules determining how to make procedural rules. That is, it consists of legislative procedures. The constitutional level defines and limits the kinds of legislation that may be adopted. Our focus is on the collective action and procedural rules. Specifically, we consider local, state, and national legislation and the effects of that legislation once it is converted into procedures and regulations.

In our analysis, we examine both institutions and organizations. We identify formal and informal organizations and attempt to discover and evaluate institutions that exist within and across those institutions. Our analysis, we believe, provides a means of better understanding the patterns of interactions that result in green goals conflicting with green policies.

ANALYTICAL CONCEPTS

Institutions and Organizations

Policy processes are messy, or more formally, they have a great deal of complexity. They operate within different levels of institutions and across organizations. They may be created by interactions that are games within games as protagonists and antagonists bargain, threaten, cooperate, or demonize. Often, groups that cooperate at one level are in opposition with each other at another level. A major reason for that complexity in the United States is that our organizational context is federalism, a system in which political power, or sovereignty, is divided among the national, state, and local governments. Distinct as well as overlapping areas of rights to govern reside in each level of our federal system. Power at each of these levels is often separated between judicial, executive, and legislative authorities. In many counties and municipalities however, the executive and legislative powers are not separated. Instead, they are combined in the county or city council. Separating power within levels of government is not required in a federal system but it is a feature of American federalism. Local political power is not restricted to cities or counties as it is also exercised by special taxing districts such as mosquito abatement or water conservation districts, associations of government and even the most rapidly increasing form of local government—homeowners' associations. All told, there are more than 80,000 different governments in the United States.

Sometimes jurisdiction between all these levels of government overlaps and sometimes it does not. The federal government can pass a law to protect endangered species that cannot be ignored by other state and local governments. Those governments can, however, establish rules more restrictive than those from the national government. California for example, has a statewide Endangered Species Act that is more powerful than the national one. Conversely, because zoning authority resides with state and local authorities, a local homeowner association may choose to establish rules prohibiting solar panels from being seen from the road. Just as the association can banish basketball hoops to the backyard, it can banish solar panels to the backside of a roof, regardless of the national government's desire to promote solar energy.

THE DEMOCRATIC PROCESS

We analyze political choices using the tools and assumptions of economics. The focus of that analysis is how individuals choose within the constraints of rules. Thus, instead of studying just the structure of bureaucracy, we also study how that structure affects the individual bureaucrat. Congress is a large, complex institution that is best understood, we believe, by studying the individual members and how they act within the formal and informal rules of Congress. Likewise, when we study interest groups, we consider the individuals in those groups. We evaluate individual benefits and costs from participating with the group. We assume that citizens, legislators, bureaucrats, and interest group members have passions and interests that they pursue through governmental processes. It is as if we view politics as a game and these various players in the game as the game pieces. Unlike a board game however, these game pieces move on their own power. They sometimes group together to form alliances to accomplish their ends. Sometimes, they sit back and watch others play.

This model of politics and the democratic process is a simple one. The players—voters, interest groups, politicians, and bureaucrats—attempt to accomplish their own ends through government. By assuming these players are self-interested, we are not assuming they are selfish. We are simply saying they attempt to accomplish their own goals. The actual goals are where notions of selfishness and otherness can be examined. We assume that many people have goals that are other-interested. Many want to save polar bears and prairie dogs, for example. Others care only for themselves. The point is that people attempt to achieve their goals even if those goals and actions conflict with those of others. Thus, our self-interest assumption is that people pursue their own goals and that those goals can be broad or narrow.

This is essentially a common sense view of politics. We all know that most of the time, people invest their scarce time in doing what they value the most. Conversely, they choose to not invest in actions that have little payoff for them. Of course, we do not always know what is best for us but neither does anyone else. All decision environments, including policy arenas, are characterized by varying degrees of uncertainty; that is, we only rarely know with confidence what will happen. This is especially true in policy decision environments. We must choose to participate, or not, in varying states of ignorance with respect to what other rational, self-interested people are contemplating. We noted above that Kiser and Ostrom identified three levels of political institutions: constitutional, collective choice, and procedural; we also noted that our analysis is at the collective choice and procedural levels. We assume that at the collective action level (Congress, legislatures, councils and boards), many players in the collective action game will attempt to have laws passed that are consistent with their own goals. Further, we expect that they will also attempt to influence the content of procedural-level rules and regulations and will work to ensure that procedural rules are followed. Citizens pursuing their private desires quickly learn that organized groups have more influence on laws and regulation setting than do individuals. They often, therefore, organize into interest groups where they can pursue their goals.

Individuals and groups lobby to have their interests reflected in the structure and outcomes of laws, rules, and regulations. Much lobbying is what economists call "'rent-seeking," which is more than just seeking favors. It is attempting to collect benefits, both physical and emotional (what economists call 'rents') from capital they do not own. The standard term used to describe them is "rent-seekers." Rent-seeking literature identifies outcomes from lobbying that are quite the opposite of that predicted by James Madison in Federalist 10. In that essay, Madison argued that the federal design provided by the proposed constitution would control the excesses of factions. He defined factions as "a number of citizens, whether amounting to a majority or a minority of the whole, who are united and actuated by some common impulse of passion or of interest, adverse to the rights of other citizens or to the permanent and aggregate interests of the community" (Madison, 1787). Madison was especially concerned about majorities taking advantage of minorities and believed that with their new form of government, they were creating a political marketplace in which majority factions would compete with and control each others' excesses. It is possible to interpret Madison's analysis as concluding that the political game played under the rules of federalism was in some ways, an exchange or gains-from-trade game.

In Federalist 10, Madison did not anticipate situations exactly opposite to majorities exploiting minorities; that is, situations in which minorities use the political system to exploit majorities. The rent seeking literature takes up that analysis by viewing political games as wealth-reducing activities. In these games, the amounts of wealth the players spend attempting to gain rents are, in total, more than the value of the rents gained. As we explain in detail below, they are locked in the tragedy of the commons. By seeking their own interests, they often make others worse off. Attempts to stop the Cape Wind project (Chapter 4) may, for example, be a form of rent seeking. If the opponents are successful in blocking the project, one likely result is that more carbon-based energy will be used than if the project were built. Rent seekers attempting to get a regulation adopted can harm other interests, as is the case of protecting the Indiana bat under the Endangered Species Act, which in effect, makes some excellent wind power sites unavailable.

Group members seldom see the costs of their actions but they do see the benefits. Prairie dog proponents rejoice at saving prairie dog colonies but they often do not pay the costs of formerly productive agricultural lands being rendered unproductive. Mandating that a city increase the amount of wind power it uses but also enacting laws that prohibit siting wind power anywhere near the city means that wind generation facilities have to be located elsewhere, usually in rural areas in other states. Thus, the city's politicians and citizens get the benefits of wind energy and the environmental and social production costs are exported to other places. A city council or state legislature can adopt a solar or wind energy mandate without considering the costs of siting and transmitting the energy. They do not have to consider the costs to a rural area of having hundreds of windmills or hundreds of acres of solar facilities, whose energy production is exported elsewhere. Likewise, groups opposing siting a wind farm on their favorite viewshed do not have to consider the costs of not reducing the amount of carbon-based energy.

Divorcing costs from benefits is what happens in the familiar setting of a group dining together and agreeing to split the bill. The situation even has a name—*the unscrupulous diner's dilemma*. If the bill is being split evenly, the selfish diner will order an exceptional dinner in the belief that her/his fellows will order normally. But, if everyone orders more in the expectation that others will help pay the cost, they each end up with a far higher bill individually and severally than if they had agreed at the outset to each pay their own portion of the bill. If friends are willing to do this to each other face-to-face sitting around a table, think how much more interests in the policy arena might be willing to do it to anonymous others.

The unscrupulous diner's dilemma was illustrated in an environmental context by biologist Garrett Hardin in his 1968 article, "The Tragedy of the Commons." He focused the attention of the environmental movement on incentives and human action. In that essay, Hardin showed that analyzing most environmental issues, from overcrowding in national parks to overgrazing commonly owned property, requires an understanding of who controls access to a resource, who gets the benefits from using it and who pays the costs. Hardin said the answers to those questions lead to basic policy principles for encouraging preservation.

In this classic article, Hardin claimed that many of our environmental problems are caused by a system of open access to a commonly owned environment. He summarized conventional wisdom about common property[1] as follows, "Ruin is the destination toward which all men rush, each pursuing his own best interest in a society which believes in the freedom of the commons. Freedom in a commons brings ruin to all" (Hardin, 1968, p. 1244). Hardin's article became one of the most cited environmental articles ever published and his call for "mutual coercion, mutually agreed upon" was the intellectual justification for decades of environmental legislation in the United States.

Hardin used a pasture as an example of how the commons can produce tragedy. As long as grazing on the commonly owned pasture is below carrying capacity, a herdsman may add another cow without negatively affecting the amount of grazing available for other cows. But once carrying capacity is reached,[2] adding the additional cow has negative consequences for all users of the common pasture. The rational herdsman faced with adding the extra cow calculates his share of the benefits (100%) and his share of the cost (1/n herdsmen) and adds the cow. And another. As do all other herdsmen. Each may care for what is common but can do nothing about it since one person exercising restraint only assures himself a smaller herd, not a stable, preserved commons. Thus, the commons is a paradox—an individual acting in his self-interest makes himself and everyone worse off in the long run but an individual acting in the group interest cannot stop the inevitable ruin.

A misunderstanding about the tragedy of the commons is the claim that the core problem is lack of conscience. If people simply developed a conservation ethic, were less greedy, less inculcated with western values and more caring of the community, the common claim is that the tragedy would not happen. Hardin rejected appeals to conscience out of hand; "To make such an appeal is to set up a selective incentive system that works toward the elimination of conscience from the race" (Hardin 1968, p. 1245). Further, to conjure up conscience "in the absence of substantial sanctions, are we not trying to browbeat a free man in the commons into acting against his own interests" (Hardin, 1968, p. 1245)? He claims, with much justification in evolutionary biology, economics, and political science, that successful policies are those that do not require people to act against their self-interest. Appeals to conscience may work in the short run but self-interest means such policies are not sustainable.

Hardin claimed the core problem to be lack of responsibility as defined by philosopher Charles Frankel. "Responsibility is the product of definite social arrangements. . . . A decision is considered responsible when the man or group that makes it has to answer for it to those who are directly or indirectly affected by it"," (Hardin, 1968, p. 1244) he said. Frankelian responsibility exists, then, when people taking an action must pay the costs of that action. Since costs also imply benefits, the other side of the responsibility coin is that the person taking the action also receives the benefits of that action.

On the commons, individuals have the authority to add an extra cow and each gains the benefits of his actions. But, the costs of each herder's actions are spread among all other users of the commons. *Any action on a commons is intrinsically irresponsible because costs are socialized and benefits are privatized.* Without the corrective feedback provided in a system establishing Frankelian responsibility, destructive actions are encouraged and, Hardin says, inevitable.

The idea of the commons is a core concept for understanding problems faced in creating alternative energy. Assuming that alternative energy makes everyone better off by reducing carbon consumption, it is clear that no one can be excluded from the benefits. But the provision dimension is another story. Alternative energy facilities need to be placed somewhere but they can be excluded from many, if not most, locations. The strategy for locals is to say they are in favor of alternative energy but the facilities need to be located elsewhere. The producers of alternative energy, thus, are restricted from accessing available sites, which reduces the amount of alternative energy production. Returning to Hardin's story of the common pasture, producing alternative energy makes society better off. But, we all have an incentive to disallow production facilities in our favorite part of the pasture. In this new story, we are not adding more cows to the commons, thus creating a tragedy through destruction. Instead, we are systematically reducing the amount of pasture available for energy production. Protecting the pasture from energy production facilities makes everyone else worse off. In such cases, small, local interests harm the general interest.

Another analytical tool consistent with and often used in conjunction with the tragedy of the commons story is 'externalities'. When people take actions that create costs for a second person without that person's permission or sometimes knowledge, they are producing what are known as negative externalities. Conversely, there are positive externalities. A family adopting a stretch of highway and keeping it clean makes everyone else better off. Their actions have positive spillover effects on others. When someone fouls the air or a stream, they are producing negative externalities.

In the context of Hardin's common pasture, adding an extra cow once the pasture has reached carrying capacity produces a negative externality. The person adding the cow gains all the benefits from adding the cow and the costs of his or her actions spill over onto all other users. The costs of adding the cow are socialized and the benefits are privatized. Socialized costs and privatized benefits mean that the person adding the cow gets a 'free-ride' at others' expense. They are free riders.

An excellent treatment of the free rider problem and its implications for group action was provided by Mancur Olson in his 1965 book, *The Logic of Collective Action: Public Goods and the Theory of Groups*. Olson and Hardin wrote about the same general issues, how individuals acting in their short-term self-interest can produce outcomes that make the group worse off. Hardin wrote about individual action in a commons and Olsen wrote about individual action (or non-action) in achieving group goals. Olson explained that the problem of groups achieving their goals comes down to the free rider problem. Everyone in the group may share a common goal but when everyone in the group gets the benefits of achieving the goal, whether they contribute to obtaining it or not, there is a strong incentive to let others work toward the goal. Or, in other words, they are incentivized to free ride.

Olson illustrated the point made by welfare economists about situations having two distinct characteristics—non-exclusive provision and non-rival-rous consumption. Non-exclusive provision means that once a benefit is provided for one member of the group, it is provided for all. That is, individuals cannot be prohibited from the benefits. Non-rivalrous consumption means that one person's consumption does not affect other peoples' consumption. Standard examples of such situations are police protection, GPS signals, flood control, and national defense. Something characterized by non-exclusive provision and non-rivalrous consumption is known as a public good. With private goods, on the other hand, usually the owner enjoys private goods, and that person's consumption means others cannot consume it. Oranges, cell phones, and automobiles are just a few from the myriad of examples.

Private goods are easily provided in markets because there are willing buyers who can only get the benefits of the private good by purchasing it. The problem with public goods is that when the benefits of something (a project or getting a new law or regulation adopted, for example) are non-exclusive and where consumption is non-rivalrous, successful group action is unlikely. If everyone thinks that way, no one produces a public good voluntarily. Public goods are, therefore, less easily provided than are private goods. The incentives are all wrong: since people obtain the benefits of a public good without paying for it. If it gets provided, there is a powerful incentive to offer nothing or little in exchange.

Consider the problem of getting people to voluntarily pay higher prices for switching their energy consumption away from fossil-based sources to greener ones. Such a switch produces a public good—lower carbon production. Everyone shares the benefits of making the switch, even those continuing to use energy produced from coal. The costs are paid by those willing to pay the higher price. Consider the problem facing a wind energy entrepreneur who hopes people will willingly pay him to produce low-carbon energy. If he tries to sell subscriptions to his service, he will find that although many people claim they want the energy they consume to be greener, few are willing to pay the higher price. Something that people want in the abstract—greener energy—fails to be produced because those same people are not willing to pay for it. Of course, if they can get others to pay for it, they are all in favor.

We note that although public goods and commons problems are similar, there is an important distinction. In the commons, consumption is rivalrous, whereas it is non-rivalrous for public goods. The distinction is important when considering policies for overcoming the free rider problem in each case. In a commons, causing people to become responsible for the externality they produce can change their behavior. For public goods, private incentives such as subsidies or other benefits can encourage a sufficient amount of people to contribute so that public goods are produced.

THE ACTION ARENA

The alternative energy game is played in the same institutional arena that traditional energy development operates in. Private interests, organized and unorganized opposition and strict legal requirements characterize that arena. Opponents to traditional energy projects discovered that the projects had to go through several political chokepoints before receiving authority to proceed. These chokepoints operate much like a firewall between a local network and the internet. Before any information can enter a local network, it must meet authentication tests, error detection tests, and rules regarding syntax and semantics before it can enter the local network. Political chokepoints work the same way. Before a project can move from proposal to actual development, it must meet the requirements established by federal legislation and regulation and, because we are in a federal system, it must also meet requirements established by state and local laws and regulations. Those proposing a project must authenticate that the proposed project will meet all applicable regulations. Because opponents are skilled at challenging a project, the proposal must not contain errors in its projections of impacts on local flora and fauna, groundwater, air, archaeological sites, sacred Native American lands or geological formations. All this has to be done according to rules and procedures created by national, state, and local legislation and court interpretations of that legislation.

Firewalls stop any information that violates or does not meet its procedures; political chokepoints do the same for proposed energy projects. Challenges to fossil fuel projects come from a broad range of sources, well organized, and funded environmental groups to local garden clubs, for example. The concerns vary. Some are simply NIMBYism (Not In My Back Yard), cloaked as environmental concerns. Others raise serious questions about human and environmental health and welfare. The choke-points allow groups to challenge project viability, environmental effects, and economic and cultural outcomes. In addition, chokepoints allow for challenges to procedures. Did the project need to meet requirements of the National Environmental Policy Act (NEPA) for an Environmental Impact Statement (EIS) or was it justified in just doing an Environmental Assessment (EA)? Did it provide a realistic range of alternatives as required under NEPA? What about Clean Water Act (CWA) or Clean Air Act (CAA) requirements? Will endangered species be affected? Did the company file the needed paperwork on time and in the proper format? Are historical and cultural sites adequately protected as required under the Antiquities Act? Were Native American tribes consulted early in the process? If the project affects any stream, does the proposal meet requirements of the Army Corps of Engineers for mitigating wetland impacts? Even if the project gets past all the federal rules, does it meet the requirements of a state land use plan or local zoning rules? Can the project be challenged using common law nuisance requirements? The list can go on and on. The point is that there are many political chokepoints and they proliferate with each new local, state, and national law, the regulations that carry out the law and relevant court decisions.

Alternative energy projects must go through the same political choke-points as traditional energy projects. Although nearly everyone favors developing alternative or "green" energy, they prefer that operations be located elsewhere. Local citizens and politicians often consider them unsightly, they may be fatal to local wildlife, legislators want to tax them to increase tax revenues, local communities want to charge impact fees to build local infra-structure, local environmental groups worry about environmental impacts and Native American Tribes are concerned with negative impacts on sites they consider sacred. Farmland preservationists worry that windmills or solar farms will destroy traditional farming. It quickly becomes clear that the constraints on alternative energy development are not just physical— long distances from transportation corridors, desert or mountainous terrain, necessary and available water supplies—but political.

The most obvious political chokepoint is the set of national environmental laws (we describe them and their requirements in some length in Chapter 3). These include the National Environmental Policy Act, the Endangered Species Act (ESA), the Clean Water Act and the Clean Air Act. Each of these acts imposes restrictions that apply to any development. A solar farm that generates as much electricity as a natural gas well and power plant takes up thousands more acres and effectively destroys all vegetation under the collectors. Wind farms disturb thousands of acres and significantly affect viewsheds. Wind turbines have also been known to kill bats and some birds, including species that are endangered. Geo-thermal plants disturb one to eight acres per megawatt (MW). Permits to disturb plant and animal life on public lands have to be granted through a drawn-out permitting process, often five years or more. At the very least, an Environmental Assessment (EA) has to be done and often an Environmental Impact Statement (EIS). Each of these involves studies, public hearings, other public input, revisions, and finally publication—all examples of political chokepoints

As noted earlier, wind and solar farms can consume thousands of acres. The visual disruption from these plants is significant. The 12 MW solar farm in Upper Sandusky, Ohio consumes eighty acres of former farmland, for example. Almost nothing grows under the solar panels because the panels intercept the sunlight. Wind turbines generating 1.5 MW typically have a hub height of 260 feet with another 115 feet from the hub to the tip of the rotor, for a total height of 375 feet. By comparison, the U.S. Capitol building is just less than 289 feet tall and the Statue of Liberty is 305 feet tall. The newest "tall" towers from GE have a hub height of 393 feet and can generate 2.3–2.7 MW. From the hub to the tip of the rotor is another 154 feet, so the structure from ground to rotor tip is 547 feet. These are massive structures. The proposed Cape Wind project in the Nantucket Sound would install 130 tall turbines across a twenty-five mile area and, although Cape Wind has been granted federal approval, local opposition is significant and legal battles continue on a variety of fronts.

State and local laws, as well as some federal rules, control the development of private lands. NEPA rules for example, do not generally apply to private lands but even projects on those lands must meet Clean Air Act and Clean Water Act requirements. They must mitigate any wetland losses and any effects on cultural or historical sites must be controlled, in addition to meeting state and local land use requirements.

Developing alternative energy on the federal estate is controlled by the national government and those lands can provide many sites for large-scale alternative energy production. After all, the federal estate contains over one-third of the nation's on-shore lands. There ought to be thousands, if not millions, of acres of public land suitable for wind and solar farms and geothermal installations. The Bureau of Land Management (BLM), for example, manages just over 258 million acres, much of it in the arid West. Surely there are many suitable alternative energy sites on just those acres, let alone on the 193 million acres of forest and grassland managed by the National Forest Service. In fact, there are many such sites. The question is not whether there are suitable lands but whether current policies will allow them to be developed.

Ideal conditions for large-scale solar generation exist on the BLM lands of the desert southwest. Those conditions include the appropriate latitude for maximizing the number of daylight hours and the geographic location's effect on weather patterns. Sites that are often cloudy are not appropriate for solar installations (Environmental Benefits of Solar Power, n.d.). In 2011, the BLM prepared a draft programmatic EIS for solar energy siting on BLM lands in Arizona, California, Colorado, Nevada, New Mexico and Utah. The BLM first identified areas with physical conditions appropriate for solar energy production that have not already been set aside as Wilderness, Wilderness Study Areas, Areas of Critical Environmental Concern, National Monuments and Parks, etc. Next, the BLM worked with the states to identify lands that could be readily developable without substantial environmental controversy. Of the nearly 99 million acres that the BLM estimates are potentially suitable for solar production, the EIS identifies twenty-four Solar Energy Zones totaling 677,384 acres, which is less than .01% of BLM lands in the five states. See Graph 1 below.

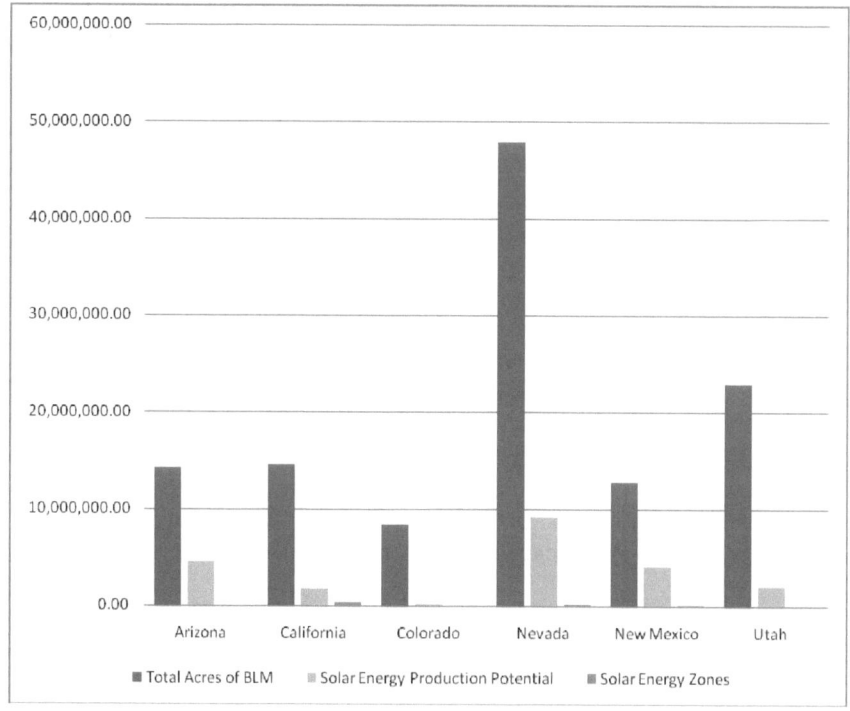

Sources:

"Public Land Ownership by State." Retrieved March 5, 2012, from http://www.nrcm.org/documents/publiclandownership.pdf.

"Solar Energy Development Draft Program Environmental Impact Statement." Solar Energy Development Program. Retrieved March 5, 2008, from http://solareis.anl.gov/documents/dpeis/index.cfm.

Figure 2.1   BLM land available for solar energy development.

Environmental groups, according to official statements at public hearings and in general, overwhelmingly support the idea of solar power and the idea behind the Solar Energy Zones in their comments at public meetings and written comments to the BLM. But, many believe the Solar Energy Zone proposals go too far, are in the wrong places and need to be modified or, in some cases, are outright rejected. Concerns were raised about the effects on the Sonoran Desert Tortoise, gullies and other riparian areas, animals and insects. The Southern Utah Wilderness Alliance (SUWA), for instance, noted that the Wah Wah Valley (one of three proposed Solar Energy Zones in Utah) is too near lands with Wilderness characteristics. Other groups complained about insufficient water in the Wah Wah Valley to sustain solar production (Chapter 5). The Sierra Club asked that two of the Solar Energy Zones in California be removed from consideration entirely. All told, the amount of 'uncontroversial' land available for solar energy production on public lands in the five states suggests that we should not expect widespread, significant production of solar from the public lands in the near future (Solar Energy Development Draft PEIS, n.d. ; Solar Energy Development Draft Program Environmental Impact Statement, n.d.).

One of the serious chokepoints for developing any form of alternative energy on the public lands is the time and energy that must go into producing and reviewing an Environmental Assessment or Environmental Impact Statement. A President may order agencies to expedite review of an EIS or EA (as President Bush did to speed up the approval process for oil and gas and President Obama did to speed up approval of solar and wind-generating facilities) but ordering an agency to expedite something and actually having it happen is not the same thing. According to our conversations with BLM officials, even expedited reviews in the BLM take a minimum of three years and are more likely to take five years. The result is that, just as there were few actual "shovel ready" projects that could be implemented when the Congress passed the American Recovery and Reinvestment Act, there are few energy sites of any kind on the public lands that can be developed easily, quickly or perhaps ever, regardless of the wishes of a Congress or President.

There is no single action arena for approving the development of alternative energy. Approving a wind energy project, for example, often requires local building permits, changes in zoning, navigation through state regulations and production of an EIS or EA. In addition, it might require CWA Section 404 permits if it affects any of the waters of the United States. If, as is often the case in the West, transmission lines cross private and public lands, negotiations must be carried out with federal and state agencies as well as private parties. The local building board, state regulators, federal regulators, and comment periods for and challenges to an EIS and state and federal courts are separate chokepoints at which different participants challenge the project.

Progress through these different arenas or chokepoints is not necessarily linear, as some arenas are in play simultaneously. Others require approval in another arena to even be considered. A Section 404 permit for example, is only considered after an EIS is completed. And, the Army Corps may ignore the recommended alternatives in the EIS. These processes are messy, complicated, and confusing.

CHOICES AND DECISIONS

In the spirit of the unscrupulous diner's dilemma mentioned earlier, we now turn to what we call the green vs. green dilemma. The dilemma is as follows: green activists sit down at the alternative energy development policy table. If the individual pays the costs of each person's choosing to preserve a local area from alternative energy development, those costs will be weighed against the local benefits of preservation. But if everyone at the table shares the costs of preservation, while the benefits remain local, there is little incentive to compare total costs and benefits. Just like each diner ordering the expensive meal, each activist chooses local preservation. The result is lots of local preservation but little alternative energy development. Activists usually base their opposition to new development, especially on public lands, on the value they place on a particular area such as a viewshed, watershed or on local fauna and flora, some of which may be endangered. Clearly, people are attracted to wild places for psychological reasons and development would harm the enjoyment people get from them remaining wild. Local environmental resources such as watersheds and endangered species however, are more than amenities. They are valued parts of integrated or interwoven natural systems and ecosystems. They provide ecosystem benefits for which there are few metrics for determining their value to humans.

What we have is an action arena in which some participants place essentially infinite value on a local resource that could be substantially harmed by alternative energy development. These participants strongly object to being classified as Not In My Backyarders. They see themselves as trying to 'save' their local environment, not as serving some private interest. NIMBY, as generally defined, is pejorative as is its companion term, BANANA (Build Absolutely Nothing Anywhere Anything (or Anyone)). A reason that activists react negatively to being labeled as NIMBYs or BANANAs is that the terms are usually construed as referring to people acting out of narrow, selfish interests. Activists we have talked to genuinely believe they are acting out of generous and even altruistic interests.

We believe that NIMBY is not an accurate way to evaluate the green vs. green dilemma. Local activists believe they are acting to protect local but broad interests, not narrow ones. Saving habitat for a local endangered species or preserving a viewshed is acting to preserve something to which the activists attach very high, if not infinite value. On the other hand, their share of the benefits from increased amounts of alternative energy is not infinite or even large. If increasing alternative energy production is a public good, it is a relatively low-valued one. We are back to the free rider problem in which locals say they support increased amounts of alternative energy but the local contribution will be inconsequential to the overall goal. They decide, rationally, to protect their local resources.

One way to view the dilemma and the decision to act or not when a alternative energy project is proposed is in the form of a decision tree. The decision to fight has a potential payoff of infinite value and the cost of the action is shared with everyone else in the country. Even a small chance of winning is sufficient to motivate. The decision to not fight is often chosen if the development has an infinitely negative payoff and a fractional positive payoff.

As is evident in Fgure 2.2, local interests will dominate the decision calculus. Of course, there is still the free rider problem as one voice is not likely to be decisive, so one gets the benefits or costs of an action whether or not s/he participates. But, environmental activists are likely to have a higher degree of 'subjective efficacy' than others (Moe, 1988; Acevedo & Kreuger, 2004). This trait, which leads them to overestimate the significance of their own contributions, encourages them to join groups. A study of participation in Common Cause, a non-profit citizen watchdog group, showed that a combination of low cost and laudatory purpose encourages people's initial decision to join (Rothenberg, 1992). Those most likely to join Common Cause were affluent, highly educated, politically active and politically liberal—all traits that were sufficient to overcome a low initial cost of membership. In addition, those who remained members stayed because of a 'learned commitment' to purposes i.e. they found the group's purposes compelling and the activities pleasurable. This suggests that at least some people will overcome the urge to free ride and will involve themselves in protecting local environmental assets. These activists show up to public hearings, try to get appointed to stakeholder forums and submit testimony and even their own studies about the proposed project. Such ongoing participation may not stop a project but it might change it enough to make it more palatable.

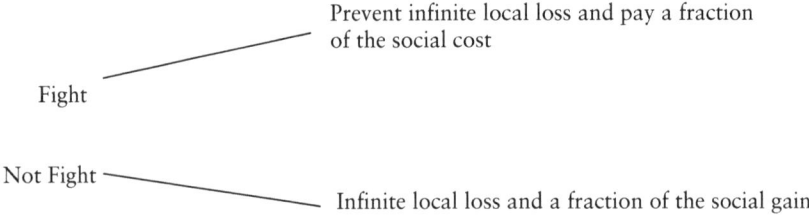

Figure 2.2 Decision to fight.

The decision to not fight may be only temporary. An activist may consider in which action arena he or she is most likely to be effective. Participating in public hearings and stakeholder forums is something that many project opponents avoid. Instead, they move to a completely different arena—the courts. One interest group that has become a frequent guest of the court system is WildEarth Guardians (WEG). WEG has filed scores of suits against those they feel have violated environmental laws. They have petitioned to have thousands of rivers in Colorado protected from any intervention, including hydroelectric projects. Any instances where they feel companies are abusing the environment and thus neglecting or at best, not aware of environmental laws, they take those companies to court to ensure they are aware of their actions.

The decisions activists face in the various action arenas can be categorized according to three choices identified by A. O. Hirschman's (1970) book, *Exit, Voice and Loyalty: Responses to Decline in Firms, Organizations and* States. When faced with a decline in one's benefits, Hirschman said a person could *exit* the relationship, *voice* opposition and provide proposals for change or remain loyal to the situation. Loyalty to firms or brands is an example. Accepting a policy based on party loyalties or because it comes from a politician one attaches him or herself to are others. It may be too expensive for an activist to participate in each action arena that exists in the American system of federalism—local planning board, city or county council, agency hearings and court proceedings, for example. A rational strategy, once one decides to not exit all action arenas, is to be loyal in some and exercise voice in others.

Table 2.1   Exit, Voice, and Loyalty

Choice	Outcomes
Exit	Physically or emotionally leave the action arena. Those who exit from one arena, however, may return to the issue when it enters another action arena.
Voice	Be actively involved in opposing the project. Submit testimony at public hearings, get appointed to stakeholder forums, and submit studies relevant to the process and issue. Organize opposition and post propaganda online. Meet with politicians. Stir public opinion.
Loyalty	Accept the process and believe that it will produce a 'fair' result. Alternatively, believe that one person's participation has little value and invest time and energy into something in which s/he is more likely to have an impact.

Source: Hirschman (1970).

INTEGRATING THE ANALYSIS

In order to understand why local green groups chose to oppose regional alternative energy projects, it is necessary to view their decision through the IAD lens. It seems almost unorthodox, at times, for conversationalists to oppose projects that would ultimately have a positive effect on the environment and would lead to less pollution in the long run. It is the decision by energy companies to site their project in what the green group views as an *especially* damaging location that often riles the conservation group; they see a myriad of other options for siting. Even if on a personal level, individuals that are part of a conservation group think that employing more alternative energy is preferable to less. As part of that group, they will have faith in the process and might even begin to actively oppose alternative energy plants.

Policy Arena

Since all actors in the following case studies are operating in the same policy arena, that context must be thoroughly understood. One of the major conflicts that will characterize these studies is the inherent friction created from the federal system. As will be explored in depth in the following chapter, many federal environmental regulations are written so as to ensure that the federal government is able to retain exclusive authority over their lands, despite their location. Thus, states are forced to accept a certain level of interference. States can, however, choose to make laws that are stricter than those imposed nationwide. For example, Utah has a higher smoking age -nineteen- than the rest of the country, where individuals can smoke once they are eighteen. Understanding the policy arena also incorporates an understanding of alternative energy type, siting conditions and specific area conditions, all of which are factors explaining why that particular area was chosen. Additionally, there should be an exploration of the preferred policy outcomes and how those match up with the actual outcomes, if they even match up at all.

Table 2.4   Levels of Stakeholders

Local Stakeholders	State Stakeholders	National Stakeholders
Citizens, user groups	User groups	User groups
Local politicians	State politicians	National politicians
Homeowner associations, special districts	Public Service Commission	Rent-seekers
Energy companies	Utility companies	Energy companies
Green groups and nonprofits	Green groups and nonprofits	Green groups and nonprofits
Planning and zoning commissions	State regulators and the judiciary	National regulators and federal courts
Farmers and ranchers		

Stakeholders

Once the context is understood, the agents acting in that context must be considered. Generally there are several key stakeholders, those at the local, state, and federal level. The beliefs, values, and preferences of these stakeholders must be reviewed and compared with their respective communities. Are these stakeholders benefitting from the proposed energy plant or what do they stand to lose if the plant is or isn't sited? What is the preferred outcome of the situation for each stakeholder, how do these outcomes match up with the realized outcome and how much control do these stakeholders have over the outcome? The positions of each of the players must also be considered, in addition the information available to them and how they use that information.

Generally, these players fall into four distinct categories. First, we will discuss citizens. These citizens may be user groups that enjoy access to these areas and the beauty or recreation it provides them. Second, are the politicians. Politicians will generally push for policies that got them elected to office and will keep them in office. Also, politicians have the power to appoint those in the third category—bureaucrats and regulators. It is those in charge of the regulating agencies that decide how legislation passed by politicians is implemented and applied. These bureaucrats implement laws at every level of government, from small homeowners' associations regulating the use of solar panels to national agencies managing wind energy proposals. Finally, interest groups can lobby politicians, bureaucrats, and citizens for their support. These interest groups may be on other sides of the issue and will lobby for outcomes that benefit them most at every level of government.

Each of these stakeholder types has a set of tools available to them; some tools are distinctly assigned based on stakeholder type, other tools are available to all types, regardless. Citizens can petition interest groups, donate money to causes, or join groups in order to expand their impact. Both citizens and interest groups share many tools in common. Citizens and interest groups can take advantage of politicians, especially during an election year. If they do not receive the answer they want from a politician, they can start a voter initiative. As part of these groups or as an individual, citizens can file lawsuits in order to have energy plant construction halted or ended all together. When proposals are put up for public comment, citizens and groups can use this period as a way of creating noise around a project proposal so it will receive more attention. Additionally, citizens and groups can protest projects or make themselves into an unofficial watchdog.

Politicians have several tools that are unique to their position. They can propose and pass legislation that will either help or inhibit projects. Under their legislative duties, politicians also have the power to implement taxes or fees and set mandates and goals that will affect projects. In order to gain support, politicians can make policy statements regarding their opinion on a certain problem or how they would like to see an outcome be achieved. They appoint bureaucrats and individuals at regulatory agencies. Politicians also control the funding that these agencies receive, which can in turn, dictate how powerful these agencies are.

Bureaucrats and members of regulatory agencies set and control many of the barriers to entry for energy projects. Homeowners' associations regulate covenants, conditions, and restrictions. Lower level bureaucrats implement nuisance laws, planning and zoning laws and set local procedures. Utility company bureaucrats set rates and make decisions over what type of energy they choose to buy. Further up the bureaucratic chain are policy implementers; here federal and state environmental institutions are created and framed.

All of these stakeholders have access to the following set of tools, some of whom have learned to use them expertly. Lobbying politicians is available to all stakeholders, even politicians, who can pressure each other to pass legislation. Stakeholders can also take issues to court and make the judiciary clarify legislation. Using media to change public opinion and garner support for your cause is a widely available tool, which must be used carefully. Social media has become a newer tool, and a favorite of conservation groups.

Table 2.5 Stakeholders' Tools

	Citizens	Interest Groups	Politicians	Bureaucrats and Regulators
Voice: Clamor, media, letter writing, initiatives	X	X	X	
Procedural Tools: Comments EIS/EA, zoning, studies	X	X	X	
Lobbying	X	X	X	X
Policy Tools: Rate setting, taxes, mandates, regulations			X	X
Courts: Nuisance laws, procedure	X	X		

Rules in Use

The next step is to analyze the rules that stakeholders adhere to. Rules may be formal listed rules such as legislation, or informal rules such as implicit community standards. Formal laws include all of the federal environmental institutions that have multiplied over the past fifty years. Any federal laws that govern businesses may also be part of these formal rules. In this text however, we focus only on energy siting. States also have their own set of formal rules. Some states such as Massachusetts and California have their own NEPA process, an environmental review procedure that will be explored in depth in the next chapter. Local counties, cities, districts, and even homeowners' associations may also have their rules regarding how energy can be sited and how the construction of these sites can occur.

Informal rules involve the unspoken set of cultural values and societal norms that govern how people act (Smajgl, Leitch, & Lynam, 2009). These rules can include how individuals are expected to operate once they join an interest group or a regulatory agency. The way in which the four types of stakeholders interact is also governed by informal rules. Many of the rules describing how these groups should use the tools available to them such as the media, are governed by these unspoken standards. Also, using these rules in a more untraditional way than what has been described above, outside of their customary use, fits into this description.

Analyzing Outcomes

Once all of the above components are understood, then the final outcome of the situation can be analyzed. First, all of the possible outcomes should be recognized. Then, achieved outcomes should be compared with the policy objectives. It is important to note whether the results were satisfactory or important and if they had a lasting effect. What are the expected results of these outcomes and will they affect areas outside of the immediate policy arena?

In order to fully understand the stakeholders, their expected or preferred outcomes should be studied. Citizens will prefer whichever outcome will give them the most utility and benefit; this may mean cheaper utility rates, more energy options or greener energy options. Interest groups will want to see their policies being implemented. Politicians will seek after rents or votes, possibly both at the same time. Bureaucrats will try to seek outcomes that will allow them to keep their jobs, meaning they will trend towards outcomes that have little negative impact or controversy.

The purpose of the following narratives is to analyze the patterns of interaction between green interest groups, energy siting companies, regulators, and regulating agencies, citizens, and politicians. By understanding the patterns of interactions, readers will be able more thoroughly understand the outcomes and why those outcomes matter. Additionally, readers will be able to understand how to apply the IAD framework in order to understand future green v. green conflicts.

3 Regulations

In order to understand the current state of alternative energy development, we examine how the regulatory institutions that attempted to protect the environment have proliferated. The environmental laws we discuss in this chapter have been created to serve several purposes. Their purpose is to protect federal land holdings, historical places and symbols, address concerns over pollution and the safety of flora, fauna and humans, and even to serve as reparations for the historical mistreatment of Native Americans. In addition to the many federal regulations that shape environmental institutions, governments use subsidies and tax policy to influence the alternative energy market.

PROTECTING FEDERAL LAND HOLDINGS

One of the very first environmental laws enacted in the United States was the 1872 General Mining Act. As miners began to move out west in search of gold, arguments arose over access to minerals on public lands. Public lands were frequently used as open mines in the rush to extract ore and other minerals. In order to preserve their land holdings Congress codified a system for staking out mining claims and protecting federally owned land (Baker, Macris & Patterson, 2002). The unseen consequence of this bill, however, was the flurry of mining and oil claims that came following the enactment of this bill (Hays, 1999). The onslaught of claims was so overwhelming that it inspired President Taft to sign an Executive Order to allocate three million acres of land for conservation (*United States v. Midwest Oil Co.,* 1915). Congress in turn passed the Mineral Leasing Act in 1920, clarifying lease qualifications and setting maximum acre limits those companies could lease (Mineral Leasing Act of 1920 as Amended). It put the responsibility for these leases in the hands of the Bureau of Land Management (BLM) and the Department of the Interior (Mineral Leasing Act of 1920 as Amended).

One effect of the General Mining Act of 1872 was to enlarge the Department of the Interior's responsibilities and authority. The Secretary of the Interior was now in control of allowing fuel pipelines across federal lands where it isn't already prohibited (Mineral Lands and Mining of 1995, 30 U.S.C. 185). These responsibilities were further expanded in 1920 with the Mineral Leasing Act, when the BLM was put in charge of more minerals, coal, and phosphate mines (Mineral Leasing Act of 1920 as Amended). The Mineral Leasing Act was amended with the Mineral Leasing Act for Acquired Lands (MLAAL) in 1947. Here the BLM was given jurisdiction over almost 700 million acres of land. Additionally, they were given charge over the development of any coal resources found on these lands (Coal, n.d.).

An amendment to the Mineral Leasing Act (MLA) was the Geothermal Steam Act of 1970, in which changes were made to the way that the BLM could administer leases for geothermal sites. It also made some changes to how royalties and revenues were collected from those areas (Geothermal Resources, n.d.).

All of the above mentioned mining acts ensure that the federal government retains authority over its land holdings. President Taft's move to conserve three million acres of federal land ensured that the government would be able to have continual authority over that land. While this move was done to protect this land from perceived overuse, it had the additional political benefit of allowing federal agencies to continue their administration over that land. Further mining acts also worked towards the goal of conservation, while keeping control over the land at the federal, versus the state level.

The Antiquities Act has become one of the more controversial of the environmental protection laws. Created in 1906 and approved by Theodore Roosevelt, the Act allows the President to grant national monument status to certain public lands through an executive order. Thus, any president has the authority to arbitrarily prevent all future development on public land (Coast Guard, 14 USC § 431, 1915). The intention was to allow the president to set aside small areas with stunning geographic features such as the Grand Canyon and Devils Tower (NPS Archeology Program, 2011). There is fear, however, that this law could be used to permanently prevent development including roads, pipelines, and even some forms of recreation, thus removing the possibility of a state or city exercising ownership over its land entirely.

Concerns over human incursion, including vehicles and permanent structures on federal lands, led Congress to create and pass the Wilderness Act. It took eight years and sixty-six revisions, but Congress was finally able to pass this piece of legislation in 1964 (Wilderness Act of 1964, The, n.d.). The Wilderness Act defines wilderness as an area where "the earth and its community of life are untrammeled by man, where man himself is a visitor who does not remain" (Wilderness Act of 1964, The, n.d.). Congress categorized 9.1 million acres of land as wilderness to "assure that an increasing population, accompanied by expanding settlement and growing mechanization, does not occupy and modify all areas within the United States and its possessions, leaving no lands designated for preservation and protection in their natural condition." (1964 Wilderness Act, 2008). One of the most important features of this legislation was the transfer of power over wilderness decisions from agencies, such as Fish and Wildlife, to Congress, thus allowing this legislative body more power over decisions made regarding federal land (1964 Wilderness Act, 2008).

The definition of protection under the Wilderness Act excludes energy production in protected areas. Controversy exists over what can and cannot be done in areas recommended for protection that Congress has not yet designated for preservation. In early 2010, seventy-two Congressional Democrats sent a letter to the Forest Service suggesting that motorized vehicles be prohibited in wilderness study areas. Responding to the letter, House Republicans sent a message to the Chief of the Forest Service in April of the same year arguing that doing so would turn areas into "de-facto wilderness" and prohibit positive development, such as energy production (Ring, 2010).

In order to encourage the conservation of some federal lands Congress established the Land and Water Conservation Fund in 1965. Money for this fund is collected from offshore oil and gas leases, with a cap of $900 million annually (LCWF Purchases, 2011). This money is to be used for procurement of land in local conservation efforts as well as for the protection of national parks, forests, and wildlife areas (LCWF Purchases, 2011). Federal, state, and local governments can submit projects to the Strategic Landscape Acquisition Ranking System. A panel then scores these projects and they are sent a committee from the Office of Management and Budget in the Department of Agriculture (LCWF Purchases, 2011). Once this committee picks a project, they match up to 50% of the submitted government funding (Land & Water Conservation Fund, n.d.). Environmental activist groups support the LWCF, and are critical only in that they believe it is not enough. The Wilderness Society claims that the *need* of funds for federal acquisition of land is around $27 billion, while the cap on the LWCF remains at $900 million (True Grit, 2000).

After concerns that the federal government was selling land for less than some deemed it worth, the Federal Land Policy and Management Act was passed in 1976. It is often called the "BLM Organic Act," and it governs the way the Bureau of Land Management manages public land (Federal Land Policy Management Act of 1976, 2001, p. 59). FLPMA declared that public domain lands would be retained under federal ownership unless disposal of a particular parcel served the national interest, but does not exclude private interests from developing and using the resources on public lands (Federal Land Policy Management Act of 1976, 2001, p. 6). The law also specified that the US government should receive fair market value for use of public lands (Federal Land Policy Management Act of 1976, 2001, p. 1). Finally, it requires the BLM to manage for "multiple use and sustained yield" (Federal

Land Policy Management Act of 1976, 2001, p. 20). In 2000 and 2010, the legislation was modified (P.L. 106–248), so that the BLM can now sell land, and use the revenue to purchase inholdings and other lands from willing sellers (Federal Land Policy Management Act, 2010, p. 1).

Because FLPMA covers most of the activity involving the BLM and land leases, it is generally included in most lawsuits and controversies. Prior to the FLPMA, the president could withdraw public lands for specific uses or from sale to prevent speculation, or the development of conservation areas; FLPMA narrowed this authority. It made public lands come under the jurisdiction of The Wilderness Act, thus the president could recommend lands for wilderness protection, but only Congress could make the final decision (Federal Land Management Policy Act, 2010). FLPMA becomes problematic for alternative energy projects when the BLM uses this Act to refuse right of way (ROW) permits, which allow energy transmission across BLM land.

The National Forest Service's land use policies also came under fire in the 1970's sparking the Forest and Rangeland Renewable Resources Planning Act of 1974, which was amended and renamed the National Forest Management Act (NFMA) two years later (National Forest Management Act, n.d.). This act originated from controversial clear cutting policies in the Bitterroot National Forest in Montana, earlier that decade (The Environmentalist and Public Participation Era, 2008). The Secretary of Agriculture is required to assess forestlands, develop a management program based on multiple-use, sustained-yield principles, and implement a resource management plan for each unit of the National Forest System (National Forest Management Act, n.d.). Additionally the Act states that it is in the nation's interest for the Forest Service to assess the nation's public and private renewable resources and develop a national renewable resource program (National Forest Management Act, n.d.). The Act also appropriates funding ($200,000,000 annually) for a Renewable Resource Program run by the Secretary of Agriculture, to be governed under the principles contained in NEPA and the Multiple-use Sustained-Yield Act of 1960 (National Forest Management Act, n.d.).

Conservation groups have brought cases against the National Forest Service for not implementing the law correctly, or for not using the correct scientific procedures. In 1995, for instance, the Sierra Club sued the NFS over its decision to allow timber cutting in part of a Wisconsin National Forest (Sierra Club v. Marita, n.d.). The Sierra Club claimed that the NFS was using "junk science" by only monitoring indicator species and not the entire biodiversity system. They further argue that if the NFS was going to give sections of the Forest to timber companies it was violating the NFMA. The courts ultimately sided with the NFS, saying the science they used was adequate (Sierra Club v. Marita, n.d.).

The Federal Cave Resources Protection Act was passed in 1988 as an effort to ensure the conservation of caves and caverns on public land. In the Act, Congress declared that caves are "an invaluable and irreplaceable part of the Nation's natural heritage" as well as an important factor in local ecosystems (The Federal Cave Resources Protection Act of 1988, n.d.). To reduce threats caused by improper use, recreational demand, urban spread, and lack of statutory protection, the Act instituted federal conservation of significant caves on public lands by prohibiting certain actions. Fines and terms of imprisonment were put in place for anyone who was found guilty of having disturbed, harmed, defaced, or altered the cave, as well as altering the free movement of animal and plant life (The Federal Cave Resources Protection Act of 1988, n.d.). Fines were also imposed on any who engaged in the bartering of goods from federally protected caves or anyone who sponsored prohibited actions (The National Speleological Society, n.d.). The Federal Cave Resources Protection Act is cited as one of many laws that regulate the development of wind power (Rothstein, 2006).

The most recent of the federal government's attempts to keep authority of their landholdings is the Federal Lands Recreation Enhancement Act (FLREA). FLREA was passed on December 8, 2004 as a part of the broader Consolidated Appropriations Act. It gives authority to government bureaus and agencies to charge a fee for recreational use and apply those fees to the maintenance and upkeep of the federally controlled lands. "Maintenance" in this Act includes providing and preserving visitor services and recreation facilities such as "trail maintenance, toilet facilities, boat ramps, and interpretive signs or programs," while specifically stipulating that revenue not be used for monitoring of threatened and endangered species (Federal Lands Recreation Enhancement Act, n.d.).

The FLREA expands the fee authority granted under the 1908 Land and Water Conservation Fund Act and the 1996 Recreation Fee Demonstration Program. The FLREA gives "fee authority" to the Bureau of Land Management, Bureau of Reclamation, Fish and Wildlife Service, National Park Service, and the Forest Service, while differentiating between the types of fees each agency is allowed to charge (Federal Lands Recreation Enhancement Act, n.d.). Fee types include: entrance fees, standard amenity fees, expanded amenity fees, and special recreation permit fees (Federal Lands Recreation Enhancement Act, n.d.).

The passage of the act was met with controversy, and "riled elected officials and environmental and recreational groups across the West" (Eskridge, 2008). Prominent among opponents was the Western Slope No-Fee Coalition, a conglomeration of groups representing a broad range of interests, from the Montana Logging Association to the Sierra Club. In addition to interest groups, state legislatures in Montana, Colorado, Oregon, Idaho, and Alaska all passed resolutions calling for a repeal of the FLREA. Critics of the Act labelled it a "Recreational Access Tax" and argued that the Act created a motivation for regulators and bureaucrats to increase prices for access to public lands. In addition, skeptics of the Act questioned the procedure by which it had been passed into law, as an attachment to a larger spending bill instead of on its own, meaning that Congress had not voted on the bill specifically (Faehner, 2006). The FLREA remains controversial as groups such as the Western Slope No-Fee Coalition continue to call to for its repeal.

PROTECTING FEDERAL WATERWAYS

President Lyndon B Johnson signed the Wild and Scenic Rivers Act (WSRA) into law on October 2, 1968. In this act, Congress proposed to preserve the free-flowing nature of rivers possessing "outstanding natural, cultural, and recreational values." Since its inception, over two-hundred rivers in thirty-eight states have been protected (Wild and Scenic Rivers). The Act's protection of rivers from projects that could alter their free-flowing nature obviously excludes many possibilities of hydroelectric power development from regulated areas. WSRA has also been cited as one of many laws that could potentially regulate both wind and solar energy development projects (Rothstein, 2006; Laws and Regulations Applicable to Solar Energy Development, n.d.).

In 1974 Congress passed the Coastal Zone Management Act (CZMA) in response to growing development in coastal areas. The direct policy implemented in the act was to preserve, protect, enhance, and restore the coastal areas of the United States, including the shores of the Great Lakes (Congressional Action to Help Manage Our Nation's Coasts, 2011). Authority was vested in the Office of Coastal and Resource Management (OCRM), a division within the National Oceanic and Atmospheric Administration (NOAA), to manage coastal resources and make the vital decisions of how to balance economic development with conservation efforts (Coastal Zone Management Act of 1972, 2011).

CZMA outlines two national programs to be put into effect, the National Coastal Zone Management Program, and the National Estuarine Research Reserve System. Currently there are thirty-four related programs nationwide governed by CZMA (Congressional Action to Help Manage Our Nation's Coasts, 2011). CZMA is applicable in regulating energy development in coastal areas, and could potentially affect biomass, carbon sequestration, geothermal, hydro, solar, and wind power development (Laws and Regulations: Coastal Zone Management Act, n.d.).

PROTECTING HISTORICAL SITES AND SYMBOLS

From 1935–1940 Congress passed several acts attempting to preserve historical sites and traditional symbols such as the Bald Eagle. Following the precedent set with the Antiquities Act, Congress passed the Historic Sites Act (HSA) in 1935 (Historic Sites Act of 1935, 1935). It was with this Act that Congress declared, "it is a national policy to preserve for public use historic sites, buildings and objects of national significance" (Historic Sites Act of 1935, 1935). The HSA puts the Director of the Interior in charge of these historic places. The Director also has the duty to find other historical sites to protect, and acquire lands that should be protected because of their historical value. HSA also established the National Park System Advisory Board (National Park Service, 1935). According to the Office of Indian Energy and Economic Development (OIEED) this act could impact energy development in biomass, carbon sequestration, coal, geothermal, hydro, solar, and wind power (Laws and Regulations: Historic Sites, Buildings, and Antiquities Act, n.d.).

Congress furthered their efforts to preserve historical places with the Archaeological and Historic Preservation Act (AHPA) passed also in 1935. Under this act federal agencies are required to protect historical or archaeological data, places, or items that are under their jurisdiction (NPS Archaeology Program, 2011). Any actions that might negatively affect these items or places must be mitigated. Then in 1940 the Bald and Golden Eagle Protection Act was passed. If any actions require the transportation of either of these types of eagles, a federal permit must first be obtained (Wildlife and Fisheries, 2012).

Although Congress did not explicitly state that these above acts were part of the historical preservation movement, they paved the way for further responsibility for protecting historical sites be given to federal agencies. The Archaeological Resources Protection Act (ARPA) in 1979 and National Historic Preservation Act (NHPA) in 1966 did just that. ARPA governs how archaeological sites, situated both upon federal and Indian lands, can be excavated (Archaeological Resources Protection Act of 1979, n.d.). It was this second act—the National Historic Preservation Act—that made major changes to how these sites were governed.

The passage of NHPA greatly increased the responsibilities federal agencies had over federally funded projects, or sites with historic or archaeological significance. It created the Advisory Council on Historic Preservation (ACHP), State Historic Preservation Office and National Register of Historic Places (ACHP, 2011). Additionally the section-106 review process was created, where proposed sites are evaluated for possible negative effects. Proposed sites are to be evaluated for their impact on historic landmarks by the head of the agency responsible for the development project. After evaluation, they were to "afford the Advisory Council on Historic Preservation. . . a reasonable opportunity to comment with regard to such undertaking." (ACHP, 2011).

The ACHP has raised concerns about the potential impacts of the development of solar and wind energy on "archaeological sites, historic structures, cultural landscapes, and properties of religious and cultural significance to Indian tribes and Native Hawaiian organizations" (ACHP, 2011). This Council is not only concerned with solar and wind development sites, they even worry about the possibility of transmission lines on their properties. Since its implementation, the ACHP has seen a "significant wave" of energy projects attempt to pass the Section-106 review process (ACHP, 2011).

Continuing with the spirit of protecting symbols was the Wild Free Roaming Horse and Burro Act (WFRHBA) of 1971. It made it a crime for anyone to harass or kill feral horses or feral burros on federal land, required studies of the animals' habits and habitats, and permitted public land to be set aside for their use (Wild Free Roaming Horse & Burro Act, 2010). It also required that mustangs "be protected as living symbols of the historic and pioneer spirit of the West" and that management plans were to maintain an ecological balance in horse populations (Wild Free Roaming Horse & Burro Act, 2010). Finally, the BLM was also permitted to close public land to livestock grazing to protect feral horse/burro habitats, although they still are required to maintain land for multiple use permits.

SAFETY CONCERNS—FLORA AND FAUNA

One of the first environmental acts aimed at protecting wildlife was the Migratory Bird Treaty Act in 1918. The original intent of this bill was to protect the birds migrating across the US and Canadian border (Migratory Bird Treaty Act of 1918, n.d.). Since its ratification this law has been expanded to cover more than 800 bird species. It is illegal to

> pursue, hunt, take, capture, kill, attempt to take, capture or kill, possess, offer for sale, sell, offer to purchase, purchase, deliver for shipment, ship, cause to be shipped, deliver for transportation, transport, cause to be transported, carry, or cause to be carried by any means whatever, receive for shipment, transportation or carriage, or export, at any time, or in any manner, any migratory bird, included in the terms of this Convention . . . for the protection of migratory birds . . . or any part, nest, or egg of any such bird. (Migratory Bird Treaty Act of 1918, n.d.).

While some may scoff at the seemingly extraneous detail, the punishment for breaking this law can be a misdemeanor charge, a fine up to $15,000 and six months in prison (Migratory Bird Treaty Act of 1918, n.d.).

This law was used in July of 2010 against two oil production companies. Charged with negligence for killing birds with the exhaust from their factories, the companies fought the charges stating that the MBTA was too vague and they were unaware their actions were affecting wildlife. The 10th Circuit Court of appeals disagreed and both companies were fined (Appeals Court Quashes Migratory Bird Treaty Act Challenge, n.d.).

The involvement of US Fish and Wildlife Service (FWS) in large bodies of water came about with the Fish and Wildlife Coordination Act (FWCA) in 1934. Congress created the predecessor to FWS in 1871 over concerns over decreased fish populations (Origins of the US Fish and Wildlife Service, 2009). As concerns over fish and aviation wildlife and their respective habitats have increased, so have the areas of concern for this agency. The Fish and Wildlife Coordination Act requires that the environmental impacts towards wildlife, and their habitats and resources be given equal weight when considering development and its respective benefits (Fish and Wildlife Coordination Act, n.d.). Actions that will harm wildlife must first be brought before the FWS to see if the effects of those acts can be mitigated. Violating this act can result in fines, or imprisonment (Fish and Wildlife Coordination Act, n.d.). Because of this act FWS also has a hand in hydroelectric projects in coordination with the Federal Energy Regulatory commission (FERC).

ENDANGERED SPECIES ACT

One of the most famous pieces of environmental legislation is the Endangered Species Act of 1972. Under the ESA, certain species can be designated as threatened or endangered. Less commonly known is the section of the Act that authorizes funding for Congress to use to acquire habitats for some endangered species (Endangered Species Act, 2011). In 1973 this Act was further strengthened and an international conference was convened where eighty nations agreed to monitor and even stop the trade of some endangered and sensitive species (Endangered Species Act, 2011).

In a nutshell,

> . . .this law requires federal agencies, in consultation with the U.S. Fish and Wildlife Service and/or the NOAA Fisheries Service, to ensure that actions they authorize, fund, or carry out are not likely to jeopardize the continued existence of any listed species or result in the destruction or adverse modification of designated critical habitat of such species. The law also prohibits any action that causes a "taking" of any listed species of endangered fish or wildlife. Likewise, import, export, interstate, and foreign commerce of listed species are all generally prohibited. (Summary of the Endangered Species Act, 2011)

The ultimate goal of the ESA is to allow these species sufficient protection to continue to exist and even thrive. The ESA has proven to be a powerful tool in regulating energy. Doc Hastings (WA-04), the chairman of the House Natural Resources committee noted how the ESA hasn't lived up to its original intent.

> The purpose of the ESA is to recover endangered species—yet this is where the current law is failing—and failing badly. Of the species listed

under the ESA in the past thirty-eight years, only twenty have been declared recovered. That's a one percent recovery rate. I firmly believe that we can do better. In my opinion, one of the greatest obstacles to the success of the ESA is the way in which it has become a tool for excessive litigation. Instead of focusing on recovering endangered species, there are groups that use the ESA as a way to bring lawsuits against the government and block job-creating projects. (Excessive Endangered Species Act Litigation, 2011).

A current use of litigation under the ESA has been used to block wind farms in order to protect the Indiana Grey Bat. This bat has been on the endangered species list since 1967 and is often killed by wind turbines (Woody, 2009). One court case found that too many bats would die from the routine operation of wind turbines in an area in West Virginia. As such the wind company was encouraged to apply for an "Incidental Take Permit." This permit will essentially give the company permission to accidentally kill endangered species. This, however, takes years to get and has a rigorous application process, including the formation and application of copious habitat conservation plans (Woody, 2009). Corporations are getting in the habit more often of hiring scores of biologists to examine sites in detail to judge if they can built without harming any endangered species.

NATIONAL ENVIRONMENTAL POLICY ACT

In 1970, the most comprehensive and exhaustive environmental legislation to date was passed, the National Environmental Policy Act. Before this act, the primary means of challenging perceived environmental abuses was the common law. NEPA was based on the perception that a national policy and consistent standards were needed under which the executive branch could have more effect on environmental quality. Signed into law on January 1, 1970 by Richard Nixon, the purpose of NEPA was:

> To declare a national policy which will encourage productive and enjoyable harmony between man and his environment; to promote efforts which will prevent or eliminate damage to the environment and biosphere and stimulate the health and welfare of man; to enrich the understanding of the ecological systems and natural resources important to the Nation; and to establish a Council on Environmental Quality. (National Environmental Policy Act, 2011)

NEPA attempted to create a unified federal policy when it came to environmental procedures, and in a way secure federal dominance in ruling how people could interact with their environment. All federal agencies were now required to "adhere to certain environmental values and goals." (Luther, 2005, p. 1).

Before federal agencies or those using or leasing federal land take any action, they must first go through the NEPA process to see if their actions will have effects on the environment considered too detrimental. To get around the above qualification one can apply for a Categorical Exclusion (CE) (Protection of Environment, 40 CFR 1502, 2005). Here they must prove their actions will have no significant impact, individually or cumulatively, on the human environment (Protection of Environment, 40 CFR 1502, 2005). If their actions do not qualify for a CE an initial Environmental Assessment is performed. Meant to be a concise document, the EA is used to determine the next step to be taken, either an Environmental Impact Statement or Finding of No Significant Impact (FONSI) (Protection of Environment, 40 CFR § 1508.9, 2005).

The EA should also include a discussion of alternatives to the proposed impact, and their respective impacts. A FONSI simply explains the rationale in determining that no significant impact will be placed on the environment. If a FONSI is issued and not disputed, the process can end. If, however, an EIS must be prepared, it means a significant amount of research must be done. An EIS requires a scientific study of the anticipated environmental impact of the action, how these impacts could be avoided, and reasonable alternatives (Protection of Environment, 40 CFR § 1502, 2012). There should also be a discussion of the long and short-term consequences of the proposed action. Once each of these documents has been published they must be handed to agency responsible for the land; they make the final decision.

Compiling an EIS can take a significant amount of time; the average is 3.4 years (DeWitt & DeWitt, 2008). Each time any one of the above reports is published, the federal agencies overseeing the NEPA process are required to post the reports and allow time for public comments. Environmental groups, however, can easily extend this process by taking issues with the report and appealing for either further clarification, or an entirely new report. Due to these characteristics the NEPA process is one of the most uncertain, volatile parts of the siting process. It is important to note, that NEPA itself will never stop a project from being completed. The report can however carry much weight in the decision to allow a project to be permitted.

How to manage the Wildlife Refuge Systems has been a question since the late 1960's, resulting in the National Wildlife Refuge System Improvement Act. Signed into law by Bill Clinton in 1997, it amends and builds upon the National Wildlife Refuge System Administration Act of 1966 (Digest of Federal Resource Laws, n.d.). The 1966 Act provided guidelines and directives for administration and management of all areas in the system, including "wildlife refuges, areas for the protection and conservation of fish and wildlife that are threatened with extinction, wildlife ranges, game ranges, wildlife management areas, or waterfowl production areas" (National Wildlife Refuge System Administration Act, 1966). This Act has been through five modifications, before the most current, the 1997 bill.

The most recent modification gives guidance to the Secretary of the Interior for the overall management of the Refuge System (National Wildlife Refuge System, 2010). The Act's main components include a strong and singular wildlife conservation mission for the Refuge System. The Secretary of the Interior should maintain the biological integrity, diversity, and environmental health of the Refuge System. Additionally, this bill adds a new process for determining compatible uses of refuges, including recognition that wildlife-dependent recreational uses involve hunting, fishing, wildlife observation, and photography (National Wildlife Refuge System, 2010). It also establishes that environmental education and interpretation, when determined to be compatible, are legitimate and appropriate public uses of the Refuge System (National Wildlife Refuge System, 2010). Any alternative energy organizations that would attempt to build in Wildlife Refuse Systems would be prevented from doing so due to this Act.

SAFETY CONCERNS—HUMANS

As airline travel was getting established members of Congress felt national standards and regulations were necessary. Thus the Air Commerce Act was passed in 1926. Under this law, safety standards were enforced, including rules requiring that anyone wishing to build a structure taller than 200 feet, within 20,000 feet of a military or civilian airport must notify the FAA (NPS Archaeology Program, 2011). Since most wind turbines are more than 200 feet tall, this requires the company siting the turbine to first contact the FAA. The FAA, however, does not implement any of its own environmental laws.

CLEAN AIR ACT

It was the desire by Congress to answer concerns over environmental safety and pollution and to protect federal land holdings that led to some of the most powerful environmental legislation after the first version of the Clean Air Act was passed in 1963. The CAA was the first piece of federal legislation to discuss air pollution controls. Prior to this, Congress passed the Air Pollution Control Act in 1955, however it only set aside funding for air pollution research (History of the Clean Air Act, 2012). Since its passage the CAA has been amended several times, most recently in 1990.

CAA gives authority to the Environmental Protection Agency (EPA) to develop and enforce regulations to maintain air quality nationwide. The EPA was officially created in 1970 out of several smaller offices scattered across several different federal departments. The goal was to set state-specific standards and implementation plans in order to reduce the amount of pollutants and contaminants located in the air. The 1977 and 1990 amendments were primarily designed to set new time frames and standards for the states in achieving their goals, because many areas had failed to meet deadlines for pollution reduction (Summary of the Clean Air Act, 2010). EPA's enforcement of CAA on large, stationary polluters, such as manufacturers, producers, refiners, and utilities, focuses on mandating specific output limitations or requiring installation of specified pollution control equipment. Pollution reduction on mobile sources is attempted through regulating fuel composition, mandating of fuel standards, and requiring emission-control components on vehicles (Summary of the Clean Air Act, 2010).

Environmental groups have sued many organizations and companies under the authority of the CAA. Generally these lawsuits are aimed at traditional energy producers and polluters. Rep. Ed Whitfield of Kentucky points out that these lawsuits serve to "encourage environmental law suits and aid environmental groups with fund raising," which, in turn, has led to an increase of lawsuits. These lawsuits are increasingly, and ironically, aimed at clean energy projects (Rossomando, 2011). In an exhaustive research project, Bill Kovacs, Senior Vice President for Environment, Technology and Regulatory Affairs at the U.S. Chamber of Commerce, showed over 330 such energy projects have been stalled or stopped by environmental groups in litigation over laws such as the Clean Air Act.

After concerns were raised that the Clean Air Act gave all responsibility for monitoring noise levels to the EPA, the Noise Control Act was passed in 1972. With this act states are responsible for a majority of the noise monitoring, although the federal government held the right to monitor flagrant noise abusers (Noise Control Act of 1972, 1996, p. 4). It also authorized the federal government to create national noise emission standards, conduct research into noise control and its effect on the public's health (Noise Control Act of 1972, 1996, p. 4). This Act is applicable to the construction of any type of energy site, but it can also affect the ongoing noise levels at both geothermal and wind sites.

The EPA was given more authority in 1976 with the Resource Conservation and Recovery Act. The responsibility for controlling hazardous waste from "cradle-to-grave", including the generation, transportation, treatment, storage, and disposal of hazardous waste, was given to the EPA (Summary of the Resource Conservation and Recovery Act: 42 U.S.C. § 690, n.d.). It was modified in 1980 to account for environmental problems that can result from underground tanks storing petroleum and other hazardous substances.

The Federal Hazardous and Solid Waste Amendments passed in 1984, focused on waste minimization and phasing out land disposal of hazardous waste as well as corrective action for releases. It increased enforcement authority for the EPA, created more stringent standards, and created a comprehensive underground storage tank program. Any alternative energy sites that dispose of waste through thermal destruction are regulated under this act, causing controversy as some conservationists feel this leads to more pollution.

The EPA's jurisdiction further expanded in 1976 over concerns regarding the "production import, use, and disposal" of certain toxic substances. These concerns were answered with the Toxic Substances Control Act (TSCA) (U.S. Environmental Protection Agency, 2012). Contrary to its name, TSCA does not separate chemicals into "toxic" and "nontoxic", but prohibits the manufacture or importation of chemicals that are not on the TSCA inventory or subject to an exemption. Generally, manufacturers must notify the EPA before importing or manufacturing chemicals for commercial purposes. Notable exceptions include chemicals for research and development purposes. If the agency finds an unreasonable risk to health or the environment, the agency may regulate, limit uses or production volume, or even ban a substance. A variety of environmental agencies have protested this act because they feel it is not thorough enough (Earth Talk, 2011, p. 1).

After putting the responsibility for managing the importation of toxic substances under the EPA, Congress expanded this power in 1980 with the Comprehensive Environmental Response, Compensation, and Liability Act (CERCLA) (CERCLA Overview, 2011). Commonly known as the "Superfund", CERCLA is designed to clean up sites contaminated with hazardous substances. It grants broad authority to clean up releases/potential releases of hazardous substances that may endanger public health or the environment. It authorizes the EPA to identify responsible parties and compel them to clean contaminated sites, and sets up a trust fund (created by a tax on chemical and petroleum industries) to fund EPA efforts to clean sites when no guilty party can be found (CERCLA Overview, 2011). Additionally, it created a National Priorities List, and revamped the National Contingency Plan, which are response plans in case there is a release of hazardous substances (CERCLA Overview, 2011).

In 1990 the EPA's jurisdiction was once again expanded with the Pollution Prevention Act. The aim is to lower pollution levels by forcing cost-effective changes "in production, operation, and raw materials use." (42 U.S.C. §1310, 2012). Growing out of the Toxic Substances Control Act, regulators hoped that by preventing pollution they'd be able to lower pollution levels. According to the EPA the purpose of this law was to move up the chain to try to limit waste at its source (Browner, 1993).

CLEAN WATER ACT

Originating in 1972 as the Federal Water Pollution Control Act, the Clean Water Act has left a sour taste in the mouths of private property owners and environmentalists alike, although for different reasons. Amended and renamed in 1977 this is the primary law governing water pollution. It establishes the basic structure for regulating discharges of pollutants into the waters of the United States and regulating quality standards for surface waters, all of which are regulated by the EPA (Water Pollution Control Act of 1972).

The CWA is involved in quite a bit of regulation and case law against industrial pollution, usually in hydroelectric regulation conflicts. A primary example can be found in the case S.D. Warren Co. v. Maine Board of Environmental Protection. S.D. Warren Co. operates several hydroelectric dams in Maine, and they sued over being required by the Federal Energy Regulatory Commission to receive approval from the Maine Board of Environmental Protection (*S.D. Warren Co. v. Main Board of Environmental Protection*, 2011).

The Clean Water Act has also attracted criticism from conservation biologists. Sediment is classified as a pollutant under the CWA, but high amounts of sediment are crucial in some rivers because of the geological formations they create, and because they create and maintain deltas and wetlands, where a variety of endangered species live (Zellmer, 2011, p. 89). In particular, Section 404 of the CWA requires a permit from the Army Corps of Engineers in order to release fill or dredge material in waterways, and includes wetlands into the definition of waterway (Mergel, 2009). While landowners consider this an invasion of private property rights, environmentalists strongly dislike the fact the dredging wetlands is potentially legal (Mergel, 2009).

Congress increased water regulations with the Safe Drinking Water Act (SDWA), passed originally in 1974, and was amended in 1986 and 1996 (Safe Drinking Water Act, n.d.). The SDWA requires the Environmental Protection Agency (EPA) to set standards for the quality of drinking water and oversee the states, localities, and water suppliers who implement them. The SDWA covers all drinking water and sources in the United States except for private wells (used for twenty-five persons or less), and bottled water, which is overseen by the Food and Drug Administration.

The Office of Indian Energy and Economic Development lists the SDWA as one of a series of federal laws that is binding to the development of wind, solar, geothermal, hydropower, and carbon sequestration projects (Energy Resources, n.d.). The SDWA has also been cited in litigation against hydraulic fracturing, a common practice of natural gas drilling companies and the production of geothermal energy (Rogers, 2009). However, in the Federal Energy Bill of 2005, Congress amended SDWA to specifically exclude hydraulic fracturing from the stipulation against "underground injection" (Effect of Federal Safe Drinking Water Act, n.d.).

REPARATIONS

The end of the 1970's brought the era of major environmental legislation to a close. After that, it came to the attention of Congress through a series of litigations, citizen petitions, and eventually a Congressional report that several federal laws and state laws were preventing Native American from exercising their right to freedom of religion (Protecting Religious Freedom and Sacred Sites, 2008). Many of the rites used in Native American religious ceremonies are land based, thus any regulations prohibiting or restricting certain actions on federal lands can prevent worship practices. In response, Congress passed the American Indian Religious Freedom Act (AIRFA) in 1978. Any federal agencies conducting an action that might interfere with a Native American tribe's ability to practice their religious ceremonies and rites must first discuss their action with the group. The tribes included in this Act are: American Indians, Eskimos, Aleuts, and Native Hawaiians (American Indian Religious Freedom Act, n.d, p. 138).

There are many controversies involving the AIRFA. Many national Indian organizations feel that the courts do not use this act to the fullest extent intended. These groups even went as far as to protest the nomination of William G. Meyers to the head of the 9th Circuit Federal Court of Appeals because of his earlier decisions regarding the AIRFA and mining claims (Protecting Religious Freedom and Sacred Sites, 2008). There is a wish by some to even discontinue permits allowing recreation, hiking and rock climbing in their sacred sites (Protecting Religious Freedom and Sacred Sites, 2008). The Cape Wind case study illustrates how this act has been used, in part, to block the siting of a wind farm off the shores of Cape Cod.

The protections afforded in AIRFA were extended with the 1990 Native American Graves Protection and Repatriation Act. Under this act protection is afforded to Native American gravesites (Public Lands: Interior, 2012). If, in the process of construction remains are found this act defines where the remains go and the process of returning them properly (Public Lands: Interior, 2012). Depending on the site, finding these remains could stop all construction activity and the land would also be protected under the AIRFA (Public Lands: Interior, 2012).

SUBSIDIES AND TAX POLICY

To understand the arguments surrounding alternative energy, one must have a basic knowledge of federal and state policies that have promoted investment in, and development of alternative technologies. For example, federal production tax credits, categorized as corporate tax credits, currently offer a 2.2 cent per kilowatt-hour (kWh) for wind, geothermal, and closed-loop biomass production during the first ten years of production. The Department of Energy allows for unused credits to be "carried forward for up to twenty years following the year they were generated or carried back one year if the taxpayer files an amended return" (Renewable

Electricity Production Tax Credit, 2011). While some tax credits initiated by the Energy Policy Act of 1992 have been extended to 2013, wind technology's tax credits are due to expire at the end of 2012. As of December 2011, twenty-four states offered tax credits for alternative energy (DSIRE Glossary, n.d.). Proponents of similar policies contend that subsidies for alternative energy diversify the energy portfolio of the US and benefit the environment, while opponents point to, among other things, artificial increases in energy supply in the market, less competition, and a corresponding decrease in incentives for profitable, self-sustaining, innovative technologies. The Department of Energy lists thirty other various incentives and policies associated with energy in the Database of State Incentives for Renewables and Efficiency (DSIRE Glossary, n.d.). A few examples of policies that encourage development of alternative energy technology (among many others in a variety of forms) from DSIRE include renewable portfolio standards (RPS), loan programs, property-assessed clean energy financing (PACE), and public benefit funds (PBF).

Renewable portfolio standards (RPS) require utilities to produce a percentage of their electricity using either renewable energy, or renewable energy credits (RECs) (DSIRE Glossary, n.d.). The EPA describes the purpose of a RPS, "The goal of an RPS is to stimulate market and technology development so that, ultimately, clean energy will be economically competitive with conventional forms of electric power" (EPA Clean Energy-Environment Guide to Action, n.d.). Essentially, an RPS creates an artificial demand for renewable energy. Although similar to renewable portfolio goals in that both establish a schedule for reaching a particular generating capacity using renewable sources, an RPS is legally binding (DSIRE Glossary, n.d.). A "set aside," or a "carve out" is a provision in an RPS that "requires utilities to use a specific renewable resource. . .to account for a certain percentage of their retail electricity sales. . .according to a set schedule" (DSIRE Glossary, n.d.). In other words, a "carve out" is a production standard geared toward a specific technology. California established an RPS in 2002 that "requires all retail sellers of electricity to purchase 20% renewable electricity by 2010" (EPA Clean Energy-Environment Guide to Action, n.d.). DSIRE reports that twenty-nine states, the District of Columbia, and Puerto Rico have an RPS, while eight states have goals (RPS Policies, 2012).

Loan programs also provide money for clean energy development. DSIRE writes, "low-interest or zero-interest loans for energy efficiency projects are a common demand-side management (DSM) practice for electric utilities" (DSIRE Glossary, 2012). Loan rates and terms vary from one project to another, but generally last ten years or less (DSIRE Glossary, n.d.). With low-interest loans, and in some cases, loan guarantees from the federal government, renewable energy sectors have obtained financing for development. There are currently 37 states (and Puerto Rico) offering loan programs for renewables (Loan Programs for Renewable, 2011).

Property assessed clean energy (PACE) programs help finance clean energy, and efficiency improvements. The energy improvements are financed "through assessments on a property owner's own real estate tax bill" (Headen et al., n.d.). Local governments that choose PACE funding require authorization from state law (DSIRE Glossary, n.d.). Headen et al. summarize how PACE financing worked in Ohio writing, "By allowing participating property owners to pay for energy improvements to their properties via a bond issue tied to a special assessment on their property tax bill, PACE financing enables property owners to reduce energy costs with no upfront investment" (n.d.). Ohio began utilizing PACE programs in 2009 when Governor Strickland approved Ohio House Bill 1 (HB1) (Headen et al., n.d.). The legislation "allowed Ohio municipalities and townships to assist property owners with solar photovoltaic and solar thermal. . .installations through a special financing district called a 'special improvement district'" (SID) (Headen et al., n.d.). After creating SIDs, municipalities could finance specific clean technologies via "a special assessment on the real estate tax bill of any consenting, participating property owner" (Headen et al., n.d.). Additional legislation expanded Ohio's PACE program to incorporate geothermal, wind, biomass, and other energy efficiency projects (Headen et al., n.d.). There are currently twenty-eight states (and the District of Columbia) that authorize PACE financing (Property Assessed Clean Energy, 2012).

Many states developed public benefit funds (PBF) in the late 1990's to "ensure continued support for renewable energy, energy efficiency and low-income energy programs" (DSIRE Glossary, n.d.). The EPA writes, "The key objective of creating state clean energy funds with PBFs is to accelerate the development of renewable energy and CHP [combined heat and power] within a state" (Clean Energy-Environment Guide to Action, n.d., p. 21). These funds come from a surcharge, known as a "system benefits charge" (SBC) on electricity consumption (e.g., $0.002/kWh) (DSIRE Glossary, n.d.). DSIRE notes that PBF's help fund "rebate programs, loan programs, research and development, and energy education programs" (DSIRE Glossary, n.d.). Examples include Massachusetts' clean energy fund administered by the Massachusetts Technology Collaborative (MTC), and Connecticut's Clean Energy Fund administered by Connecticut Innovations Incorporated (CII) (Clean Energy-Environment Guide to Action, n.d., p. 23). Eighteen states (and D.C., Puerto Rico) have PBF's, estimated by DSIRE to amount to $7.8 billion by 2017 (Public Benefit Funds for Renewables, 2012). The EPA's Clean Energy-Environment Guide to action says that PBF-based energy funds provide a cohesive, "under one roof" strategy, can be tailored to specific states based on natural resources, state goals, and industry presence, and can support and complement long-term goals and policies such as RPSs (Clean Energy-Environment Guide to Action, n.d., p. 22). California uses PBF's to "pay the incremental cost for utility RPS compliance" (Clean Energy Environment-Guide to Action, n.d., p. 22).

CONCLUSION

The federal government has had a large role in the way humans interact with the environment. Federal legislation, whether regulatory or fiscal is expansive and controls almost every aspect of siting alternative energy plants. In many of these acts Congress asserts itself as the sole decision maker when it comes to environmental policy. The acts Congress passes reflect this, as well as desire to keep the final authority vested in their governing body.

As the following case studies will illustrate, case after case, initiated by green interest groups will use these regulations to inhibit if not prevent, alternative energy siting. In the Telephone Flats, Fourmile Hill, and Cape Wind cases arguments that these plants affected act meant to make reparations for the mistreatment of Native Americans. The ESA is currently being used to try to derail the Ivanpah Solar Plant. Generally any acts regarding rivers and waterways are used against hydroelectric plants, include the Glen Canyon Dam. In addition to these acts, the unintended consequences of regulations that promote one industry over another will be explored in the Biofuels chapter.

4 Wind Energy

THE BASICS

The Department of Energy reported in August of 2011 that "wind energy has been growing at an average rate of 25% per year, making [it] the fastest growing source of energy in the world since 1990" (Energy Basics: Wind Power Animation, 2011.). Demand for alternative energy sources and improvements in the technology behind wind power have provoked discussion from many angles. Proponents, such as the American Wind Energy Association, argue that wind is a clean, cost-effective, alternative energy source that provides security against unstable fossil fuel prices, and helps diversify the United States' energy portfolio (Utilities and Wind Power, n.d.). Besides the contentions of noise, aesthetics, and avian mortality, opponents contend that other energy sources, such as natural gas, are more cost-effective. Skeptics add that approval for wind farms is just as difficult to obtain as coal-fired plants because of NIMBY protests (Eilperin, 2012).

The Energy Information Administration describes the basic science behind wind energy in simple terms. When the sun heats the earth's surface and atmosphere at different rates, the uneven heating creates wind (Wind Energy Explained, 2011). The earth's physical geography affects wind patterns that vary according to a variety of factors such as atmospheric pressure, season, and time of day. Windmill-like machines called turbines harness wind power by converting kinetic energy into electricity. The Department of Energy provides a video describing the process, saying "Wind turbines draw upon the force of moving air to generate electricity by rotating propeller-like blades around a rotor. The motion of the rotor turns [a] drive shaft, which turns an electric generator" (Wind Energy Explained, 2011).

The make, size, and use of turbines vary. Most large, modern wind turbines are horizontal-axis turbines (resembling a traditional windmill), although vertical-axis turbines exist as well. The largest turbines have blades that "span more than the length of a football field, [stand] 20 building stories high, and [produce] enough electricity to power 1,400 homes" (Wind Energy Explained, 2011). Smaller turbines power houses or smaller structures as stand-alone or off-grid energy production sites. Wind farms are groups of large wind turbines linked into a central power plant. The electricity produced by wind farms can connect into a utility grid for distribution. Public utility companies do not own all wind power installations. Businesses called Independent Power Producers build wind installations and sell their electricity to electric utilities (Types of wind turbines, 2011).

Developers build turbines on land and offshore, siting them according to strong wind resources, wind consistency, community input, and environmental concerns. When sited properly, wind turbines harness wind energy to produce electricity. A Wind Energy Resource Map produced by the Department of Energy and the National Renewable Energy Laboratory shows the abundance of wind energy in the United States (Types of wind turbines, 2011). NREL mapped potential wind resources according to wind-power density classes, determined in part from wind speed, and measured on a scale from one (lowest wind power) to seven (highest). NREL designated the regions containing a wind power class of three or above as useable wind resources, or resources that had an annual wind speed of at least thirteen mph. According to the BLM, "Wind speed is a critical feature of wind resources, because the energy in wind is proportional to the cube of the wind speed. In other words, a stronger wind means a lot more power" (Types of wind turbines, 2011).

HISTORY

Designs for some of the first windmills were found in Leonardo da Vinci's book, *Machine Novao* (Hau, 2006, p. 13). Well before da Vinci, the Chinese were using windmills to pump water. It was not until 1972, however, that Meikle and Lee, two Scotsmen, created the first windmill. Similar to the sails on a boat, the first rotary blades were made of cloth and were 'reefed' or adjusted by hand in accordance with the weather (Hau, 2006, p. 14). Gradually the blades developed from cloth, to wood and eventually sheet metal, including auto-adjusting shutters on the blades that shifted to accommodate for increases in wind power.

Daniel Halladay invented the first wind turbine in 1854 after complaints about the constant attention settlers to the American West were forced to pay to their windmills, which were used mainly for pumping water from wells (Hau, 2006, p. 17). Halladay is reported to have said "'I can invent a self-regulating windmill that will be safe from destruction in violent windstorms, but I don't know of a single man in the world who would want one'" (Hau, 2006, p. 18). By dividing the wings into sections he was able to create a self-adjusting blade that recalibrated their pitch anytime the wind changed. Another inventor, Leonhard R. Wheeler invented the windmill that is seen more commonly on farms, using a weather vane to sense the wind's direction and a weight to change the direction of the windmill's blades (Hau, 2006, p. 20).

Harnessing wind energy to power electric technologies didn't occur until Poul La Cour began experimenting with windmills in the 1890s (Hau, 2006, p. 23). By employing the principles of electrolysis, La Cour took the direct energy current created by the windmill and created hydrogen gas, which could then be used to power local schools (Hau, 2006, p. 24). In rural areas that only had access to diesel and gas engines, wind power was a way to build power grids (Hau, 2006, p. 25). At the base of every wind-mill was an electric generator; the energy gathered from the blades was passed through this generator to small grids. Diesel and gas engines also fed these grids as they were extended to areas with large energy requirements (Hau, 2006, p. 25).

Wind power remained a mainstay of rural life in the West until FDR's New Deal introduced electric grids (How Wind Energy Works, 2009). This shift towards a traditional energy grid system in the US meant few resources were put towards advancing wind technology, and technological advancements moved back home to the Netherlands.

It wasn't until 1957 that Gedser put the first modern wind turbine to work (Hansen et al., 2000, p. 1). Using a horizontal-axis wind turbine (HAWTS) with three blades and a myriad of inexpensive, standard parts Gedser powered an (alternating current) AC-grid. Efforts to build wind turbines similar to Gedser's design increased markedly with the onset of the 1973 oil crisis (Hansen et al., 2000, p. 1). As wind farms began to sprout up, engineers began to pay specific attention to pitch control (Hansen et al., 2000, p. 1). In the 1990s a combination of government subsidies and public desire to use alternative energy boosted the attention given to wind turbine development.

Wind energy has developed significantly since Da Vinci's first designs in the late eighteenth century. According to the European Wind Energy Association, Europe's 12,000 wind turbines provide roughly 9.1% of their energy requirements, at 84 gigawatts (European Wind Energy Association, 2010). The European Environment Agency claims that, by 2020, the potential for on and off-shore wind energy will be three times greater than expected electricity demand, increasing to seven times expected demand by 2030 (European Environment Agency, 2009). China has doubled its wind energy capacity every five years, and a joint report from researchers at Harvard and Tsinghua (Qinghua) University concluded that it would be economically feasible for China to provide for all its demand for electricity through wind power until 2030 (Fairley, 2009).

While most of the industrialized world claims to be attaining significant (albeit potentially overinflated) progress, wind energy potential and development in the United States is much more lackluster. The U.S. wind industry began in California during the 1970's, when the oil shortage increased the price of electricity generated from oil. The California wind industry benefited from federal and state investment tax credits (ITC) as well as state-mandated standard utility contracts that guaranteed a satisfactory market price for wind power. By 1986, California had installed more than 1.2 GW of wind power, representing nearly 90% of global installations at that time.

Expiration of the federal ITC in 1985 and the California incentive in 1986 brought the growth of the U.S. wind energy industry to an abrupt halt in the mid-1980s. Europe took the lead in wind energy, propelled by aggressive alternative energy policies enacted between 1974 and 1985. As the global industry continued to grow into the 1990s, technological advances led to significant increases in turbine power and productivity. Turbines installed in 1998 had an average capacity seven to ten times greater than that of the 1980s turbines, and the price of wind-generated electricity dropped by nearly 80% (Utilities and Wind Power, n.d.).

To promote alternative energy systems, many states began requiring electricity suppliers to obtain a small percentage of their supply from alternative energy sources, with percentages typically increasing over time. More than twenty states have followed suit with Renewable Portfolio Standards, thereby creating what proponents call an environment for stable growth.

Even so, the United States generates only about one fourth of the total wind power that the European Union generates from wind. The average annual growth rate of wind energy in America has been sporadic, and in some years is only a fraction of the world average. For example, in years like 2007 and 2008, wind energy increased by roughly half, but in years like 2000 and 2004, wind energy increased by merely five percent. Although recent initiatives have promoted wind energy development in America, it is apparent that significant challenges remain (Logan, 2008).

Industrial wind energy generation, though a relatively new development, has its roots in the United States. A 2008 publication by the U.S. Department of Energy entitled "20% Wind Energy by 2030: Increasing Wind Energy's Contribution to U.S. Electricity Supply" provides a brief history of the U.S. wind industry (Energy Basics: Wind Power Animation, 2008). The history notes that alongside an oil shortage and increasing electricity costs, the U.S. wind industry took off during the 1970s in California with the assistance of state and federal investment tax credits (ITCs), and state-mandated standard utility contracts (Energy Basics: Wind Power Animation, 2008, p.6).

At the same time, the global wind industry took off in Europe because of "aggressive renewable energy policies" (Energy Basics: Wind Power Animation, 2008, p.6). During the late 1990s, technological advances increased the power and productivity of turbines, producing seven to ten times the amount of energy as the models of the 1980s, and decreasing the price of wind energy by 80%. The DOE writes, "By 2000, Europe had more than 12,000 MW of installed wind power, versus only 2,500 MW in the United States, and Germany became the new international leader" (Energy Basics: Wind Power Animation, 2008, p.6).

The Energy Policy Act of 1992 provided production tax credits to power producers for every unit of energy (measured in kilowatt-hours) produced from wind (Energy Energy Basics: Wind Power Animation, 2008, p.6). Consequently, as outlined in the DOE's history of the American wind industry, states like Iowa and Texas "began requiring electricity suppliers to obtain a small percentage of their supply from alternative energy sources, with percentages increasing over time." In 2005, the United States regained its position as the world leader in wind energy with the help of "increasingly supportive policies, growing interest in alternative energy, and continued improvements in wind technology and performance" (Energy Basics: Wind Power Animation, 2008, p.148).

Wind energy continues to grow. In 2010, the Energy Information Agency reported, "global wind generation increased by about 20% from 2008 to 2009, and . . . more than tripled since 2004." The United States was the leader in wind power generation for the third year in a row in 2009 (Energy in brief, n.d.). The next highest wind energy producers in descending order were Germany, Spain, China, India, the United Kingdom, France, Portugal, Denmark, and Italy. The EIA writes, "These top-ten countries accounted for more than 85% of all wind generation worldwide (Energy in brief, n.d.)."

Energy production and capacity have increased significantly. The EIA notes that even though wind generated electricity production in 2010 only accounted for 2% of total electricity produced in the U.S., it was still enough to equal "the annual electricity use of about 8.7 million households" (Today in energy, 2011). According to the AWEA, "The third quarter of 2011 saw over 1,200 megawatts (MW) of wind power capacity installed, bringing installations through the first three quarters of the year to 3,360 MW," and the "cumulative wind capacity" of the US wind industry to 43, 461 MW (Industry statistics, 2011). The industry still trails natural gas in percentage of new generating capacity, but "represents more than 20% of the world's installed wind power" (Industry Statistics, 2011).

Proponents believe that wind energy has many environmental, manufacturing, and financial benefits. Like all power plants, wind installations have a production capacity factor, defined as "the ratio of the actual energy produced in a given period, to the hypothetical maximum possible, i.e. running full time at rated power" (Wind Power: Capacity Factor, Intermittency, 2011, p.1). With a capacity factor of 20–40%, wind plants have a lower capacity factor than other forms of energy like nuclear (60–100%) and coal (70–90%) (Wind Power: Capacity Factor, Intermittency, 2011, p.1). The Renewable Energy Research Laboratory at the University of Massachusetts at Amherst claims that "it does not make sense to compare capacity factors across technologies," contending that it is more useful "to compare the cost of producing energy among the various technologies" (Wind Power: Capacity Factor, Intermittency 2011, p.2). The Renewable Energy Research Laboratory (RERL) argues that despite having lower capacity factor numbers than other power plants; wind scores higher in terms of mechanical and electrical efficiency (Wind Power: Capacity Factor, Intermittency 2011, p.3). At the same time however, RERL acknowledges some deficiencies with wind power in terms of dispatch ability, or "the ability of a power plant to be turned on quickly to a desired level of output" (Wind Power: Capacity Factor, Intermittency, 2011, p.4).

According to the AWEA "Wind energy costs have dropped over the past few years as wind turbine technology has matured, with taller towers, and with improved wind turbine efficiency" (Utilities and wind power, n.d.). As wind has evolved into a more efficient source of energy, wind proponents have touted its many advantages. First, wind energy is a clean, alternative energy source that does not emit air pollutants, toxic waste, and other pollutants inherent in other forms of energy. According to the U.S. Department of Energy, "in 1990, California's wind power plants offset the emission of more than 2.5 billion pounds of carbon dioxide, and 15 million pounds of other pollutants that would have otherwise been produced. It would take a forest of 90 million to 175 million trees to provide the same air quality." (Wind Energy Basics, n.d.). Wind turbines are sometimes located on land used for grazing or farming, allowing for multiple uses of the land, and providing ranchers and farmers another source of income (Wind Energy Basics, n.d.).

The wind industry also provides a large market for wind power installations and American manufacturing, with "over 400 manufacturing facilities across the U.S. . . . [that] make components for wind turbines" (Industry Statistics, 2011). According to the BLM, "If wind generating systems are compared with fossil-fueled systems on a 'life-cycle' cost basis (counting fuel and operating expenses for the life of the generator). . .wind costs are much more competitive with other generating technologies because there is no fuel to purchase and minimal operating expenses (Utilities and wind power, n.d.)." To a similar end the AWEA writes, "Utilities can lock in wind energy prices for twenty to thirty years because the fuel is free. They argue that this is one reason wind power has added 35% of all new generating capacity to the U.S. grid since 2007–twice what coal and nuclear added combined" (Wind Energy Basics, n.d.).

We must note that we believe many of these claims about the benefits and costs of wind energy are just wind. That is, they are pie-in-the-sky beliefs that will not meet simple market tests. But, an economic analysis of claims for wind and other energy sources is not the subject of this book. Our purpose is to show they do not meet regulatory tests.

Like other energy sources, wind has drawbacks. The U.S. Energy Information Administration notes that negative impacts of wind must be "balanced with our need for electricity and the overall lower environmental impact of using wind for energy relative to other sources of energy" (Wind Energy and Environment, 2011). The EIA listed the following drawbacks to wind energy on their website: size and visual impact of turbines on landscape, rare fires, leaking lubricating fluids, noise pollution, bird and bat deaths, and environmental impact of access roads required to maintain turbines, and the use of fossil fuels to make metals and other materials for turbines. In addition to these drawbacks, the BLM states, "Even though the cost of wind power has decreased dramatically in the past ten years, the technology requires a higher initial investment than fossil-fueled generators. Roughly 80% of the cost is the machinery, with the balance being site preparation and installation." (Wind Energy Basics, n.d.).

Another challenge with wind comes from the fact that it is intermittent and "not all winds can be harnessed to meet the timing of electricity demands" (Wind Energy Basics, n.d.). Walt Musial, principal engineer at NREL's National Wind Technology Center, identified variability and long distance transmission for use as two limiting aspects of wind technology (Kowalenko, 2011). The Energy Information Administration provides a more detailed description of the problem:

Although wind farms have relatively low operating costs, capital investment costs are significant. In addition, the intermittent nature of wind results in relatively low capacity factors, such that a wind plant will generate less electricity than a conventional thermal or hydroelectric plant of the same size and over the same period of time. As a result of the high capital costs and intermittency associated with wind, the "levelized cost of electricity" (LCOE)—or the sum of the plant's present value of capital and operating costs, divided by its generation over the plant's lifetime—tends to be higher for wind than for most conventional generation types (EIA's Energy in Brief, 2011).

Several laws regulate wind energy. For example, noise issues are among the most common sources of public opposition to wind energy development. The Noise Control Act governs the noise produced by turbines. While the federal government monitors flagrant noise abuses, states are responsible for controlling most sound emissions from energy production facilities. Another example is legislation, such as the Migratory Bird Treaty Act, and the Bald and Golden Eagle Protection Act (16 U.S.C. 703–712; Ch. 128; July 13, 1918; 40 Stat. 755). Raptor deaths at the Altamont wind facility in California illustrate the relevance of the policies protecting birds around turbine installations. Because of the height of turbines, wind developers must notify the Federal Aviation Administration–created under the Federal Aviation Act–and obtain prior approval for wind installations. Wind projects are also limited in certain areas because of endangered species' habitat, as outlined in the Endangered Species Act, and wilderness areas established in the Wilderness Act. Like other energy projects, wind installations must follow NEPA guidelines to assess the environmental impacts of a turbine installation, and alternatives.

The AWEA has identified five policies prioritized by the wind industry. These five policies include: a strong national alternative electricity standard, predictable tax policies, favorable transmission policies, and prudent siting policies.

RENEWABLE PORTFOLIO STANDARD (RPS) AND TAX POLICIES

There are several proposed policies to incentivize development of clean energies. Cap and trade, feed-in tariffs, and renewable portfolio standards (RPS) all propose different means to provide incentive for alternative energy investment. Countries like Germany and Denmark have utilized some of these methods and now lead the world in wind energy manufacturing (Winds of Change, 2010). The AWEA supports a strong national renewable portfolio standard. A renewable portfolio standard, also known as a renewable electricity standard is "a policy that sets hard targets for renewable energy in the near- and long-term to diversify our electricity supply, reduce pollution, conserve water and save consumers money" (Federal Policy, n.d.). A documentary produced by the Institute of Electrical and Electronics Engineers titled "The Winds of Change" notes that these standards place an obligation on utilities to produce a portion of their electricity from renewable sources. The Institute of Electrical and Electronic Engineers documentary outlines the RPS process as follows: first, certified renewable energy generators earn certificates for each unit of electricity they produce; second, certificates can be sold along with electricity to utilities; and third, utility companies can then pass the certificates to a regulatory body to demonstrate compliance with the standard (Winds of Change, 2010).

Similar standards are currently in place in Belgium, Chile, Italy, the UK, and dozens of states in the U.S, although there is no federal standard. Senator Klobuchar (D-MN) and Senator Udall (D-NM) both introduced a bill in the 112th Congress to create a federal standard, calling for a 25% RES by 2025 (Winds of Change, 2010).

Predictable tax policies also contribute to development in the wind industry. The Energy Policy Act of 1992 provided production tax credits of 2.2 cents/kilowatt-hour for electricity produced by utility scale wind turbines (Federal Policy, n.d.). The current PTCs will expire at the end of 2012, but Section 1603 of the American Recovery and Reinvestment Act of 2009 gives wind developers the option of receiving a 30% investment tax credit (ITC) instead of a PTC (Federal Policy, n.d.). The AWEA adds that "for projects placed in service before 2013, at which construction begins before the end of 2011, developers can elect to receive an equivalent cash payment from the Department of Treasury for the value of the 30% ITC" (Federal Policy, n.d.). Carol Tombari of the National Renewable Energy Laboratory explains that policies, like the ones previously mentioned, are necessary to "jumpstart the markets for. . .emerging technologies," and ensure cheaper energy prices in the future (Winds of Change, 2010).

TRANSMISSION

According to the Department of Energy, transmission limitations are the largest obstacle preventing the wind industry from providing a larger percentage of our energy needs (Federal Policy, n.d.). Transmission involves conducting energy from the turbines and wind farms to a central grid that distributes the electricity to consumers. Transmission is difficult with wind because areas with good wind resources are sometimes located far from grids, and the infrastructure for channeling large amounts of electricity is deficient. According to Walt Musial of NREL, "We are not short of sites to put wind turbines, we are short of transmission lines in the infrastructure to bring wind energy to the cities and places it will be used." (Winds of Change, 2010).

AWEA supports Musial's statement, specifying that "currently, almost 300,000 megawatts of proposed wind projects, more than enough to meet 20% of our electricity needs, are waiting in line to connect to the grid because there is not enough transmission capacity to carry the electricity they would produce" (Federal Policy, n.d.). On-site grid integration saves money because otherwise costs are about a million dollars a mile (Winds of Change, 2010). J. Charles Smith from Utility Wind Integration Group (UWIG) compares our current grid infrastructure deficiencies to highway problems about fifty years ago. Just like some roads could not handle a ton of traffic, Smith says, "The existing transmission system wasn't built to transport large blocks of power over long distances" (Winds of Change, 2010). The increased capacity of wind installations will require an upgraded grid, but transmission systems take longer to develop than turbines and have an even larger visual impact (Winds of Change, 2010).

SITING

Because of the size and visibility of industrial turbines, developers and landowners invest a great deal of time and energy deciding where to build wind farms, a process called "siting." Siting is not a simple procedure because one must balance many different and often competing concerns. Remote wind installations require significant upgrades in transmission lines, which are very expensive, so it is difficult to always build wind farms out of sight to appease those opposed to the visibility of huge turbines.

Beyond cost concerns, the AWEA says that prudent siting policies must "address wildlife and habitat issues, military and non-military infrastructure, and community concerns" (Federal Policy, n.d.). Inadequate siting causes many of the problems that fuel opposition to wind development. For example, inadequate siting at the Altamont wind installation in California resulted in hundreds of avian deaths, including federally protected golden eagles (Winds of Change, 2010). Additionally, siting concerns in the community impeded the development of the Cape Wind Project in Nantucket Sound. Matt Patrick, a Representative from Massachusetts' third Barnstable District said that opposition came from some of the wealthiest families and was based on "aesthetic reasons" (Winds of Change, 2010). Informed siting policies can reduce or eliminate many problems to which opponents of wind energy point such as aesthetics, noise, and wildlife deaths, but cannot ignore pragmatic concerns such as cost.

The wind industry continues to grow rapidly, fueled by political incentives, enhancements in current technology, innovation, and creative programs to fund research and investment in renewable energy. Potential growth and development is enhanced in the presence of certain scenarios developed by the Department of Energy: offshore wind development, innovative new technologies like Makani Power's *Wing 7*, and competitive pricing options from utility companies.

The 2008 publication by the U.S. Department of Energy mentioned earlier explored US-specific scenarios for reaching 20% of energy demands by 2030 using wind. The DOE developed the scenario according to several assumptions: that US electricity consumption grows to 39% from 2005 to 2030 as estimated by the EIA; that wind turbine costs decrease 10% by 2030; and that no major breakthroughs occur in wind technology. The study found that 20% wind electricity would require about 300 GW of wind generation and that affordable and accessible wind resources are available across the nation. The study also revealed that costs to integrate wind would be modest, the raw materials are available, and that transmission issues would be the biggest challenge. The direct, but incremental, cost to society would be around $43 billion (50 cents/month/household). The payoff would amount to $50-$145 billion in reduced emissions and carbon regulation costs, 8% reduction in water consumption through 2030, 150,000 direct jobs, $2 billion in local annual revenues, reductions in nationwide natural gas use of 11%, and corresponding $86-$214 billion in savings (Energy Basics: Wind Power Animation, 2008). The efforts outlined in the Department of Energy's "20% by 2030" publication shows that a scenario paralleling energy production in Denmark may prove a worthwhile investment (Energy Basics: Wind Power Animation, 2008).

Offshore wind installations are promising, but the technology still needs further development. The *Nysted Offshore Windfarm*, located in Denmark, is one of the world's largest wind farms, consisting of seventy-two turbines that "generate enough power to supply 145,000 family homes with nonpolluting energy" (Havmøllepark, 2010). Mike Robinson, Deputy Director of the National Wind Technology Center said that wind resources off the coast are excellent because of the amount of wind, higher wind speeds, and the consistency compared to winds over land (Winds of Change, 2010). Offshore wind installations must account for issues that do not exist on land. Foundational issues such as moving environments, sea level depths, and corrosion are all important factors. Offshore turbine technology will require additional advancement to reduce maintenance, transmission, and installation costs.

California's Makani Power Inc., alongside the Advance Research Projects Agency-Energy, is working on the prototype of an airborne wind turbine (Airborne Wind Turbine, n.d.). Because wind blows faster and more consistently at high altitudes, Makani Power has engineered a turbine-equipped, tethered flying wing, called *Wing 7* to access winds at higher altitudes (Airborne Wind Turbine, n.d.). The relatively small *Wing 7* (especially when compared to conventional industrial turbines) extracts energy using small rotors on the wing that power a high-speed generator as the wing sweeps through the sky in large, circular patterns. According to Makani, "the low material-intensity of. . . [their] system means less material for production, less fuel for transport, and less impact at the site" (Hau, 2006, p. 13). New technologies like *Wing 7* may potentially resolve certain issues with siting, and decrease the cost of producing electricity from wind.

One of the challenges to wind energy development involves structuring a windmill to produce maximum energy efficiency. The basic structure of a windmill has changed significantly: from rotors that required handheld adjustment, to computers that adjust the blades wind speed and direction shift. Although there are several different structures of wind turbines and categories within those types, all windmills follow a basic structure. Long blades face either vertically or horizontally and connect to the rotor. This rotor is connected to a shaft that turns the gears that power a generator, which transforms the kinetic wind energy to electrical energy. The two most common types of windmills are vertical-axis turbines (VAWTS) and horizontal-axis turbines (HAWTS).

The first type of wind turbine, the VAWTS, contains two varieties: the Darrieus and H-rotor (Eriksson, Bernhoff & Leijon, 2006, p.1420). Both look like an upside down eggbeater (Types of Wind Turbines, 2011). While this type of turbine is significantly less common, it is the same structure as some of the first windmills the Persians used around 900 AD (Eriksson, Bernhoff & Leijon, 2006, p.1420). The blades on these wind turbines run vertically, allowing wind capture from any direction, and connect to a generator at the base of the turbine. VAWTS do not require a gearbox or yaw system (a finicky and complex portion of some wind machines), because the blades are connected via direct drive, as in directly linked to the rotor turning the generator (Eriksson, Bernhoff & Leijon, 2006, p.1420). Darries turbines have curved blades and require a simple base, while H-rotor blades are straight and require and stronger foundation (Eriksson, Bernhoff & Leijon, 2006, p.1421). Because each blade on this type of wind turbine rotates in and out of the wind it experiences torque ripple, thus shortening the life of VAWTS.

The second type mentioned is generally what one imagines when they picture a windmill. Long blades connect to a gearbox that turns a generator, from which flows a cord that carries the recently changed electric energy down the tower out of the windmill to an electric grid (Types of Wind Turbines, 2011).

Because wind speed is extremely volatile, engineers are constantly inventing new means of maintaining a constant electric flow from turbines. Horizontal-axis wind turbines use a synchronous generator to sustain a constant rotor speed; this generator is also connected to the utility grid, allowing for optimized power output (Hansen & Butterfield, 1993, p. 118). Rotors are built so as to stall with variable wind speeds. As the wind speed increases the tips of the blades begin to drag, to balance the increase in wind power against the drag and maintain an unvarying power output (Hansen & Butterfield, 1993, p. 119).

These turbines are paired with electrical generators to allow for variable speed drive (VSD). VSD is appealing because it allows for increased energy capture, reduced mechanical stress, and enhanced control over reactive and active power (Hansen & Butterfield, 1993, p.119). Reactive and active power is the extra, stored power generated in the background of the respective magnetic and electric fields (The National Grid Company, 2001, p.1). By adjusting the variable drive speed engineers are able to harness this otherwise wasted energy. Other HAWTS use pitch control to adjust for varying weather conditions. As the wind speed increases, the pitch of the wind turbine is lowered.

As engineers continue to revamp wind turbines to reach optimal efficiency there are several mechanisms they focus on improving, one such component the tip speed ratio. This ratio is the change in rotor speed between the blade tip and the wind (Hansen & Butterfield, 1993, p. 119). The type of airfoil that is used to construct the blades also makes a difference in the efficiency achieved. Airfoils with high lift, low pitch movement, and low drag are preferred. It was not until the 1980s that engineers began to look at helicopters as a muse and noticed how critical the design of airfoils could be (Hansen & Butterfield, 1993, p. 120). Additionally, blade taper and twist is constantly calibrated. Another important and the most expensive component to horizontal-axis turbines is the yaw control (Hansen & Butterfield, 1993, p. 135). The yaw is the system within a wind turbine that positions the blades into the wind. Ineffective yaw systems account for much of the turbine downtime in wind parks and thus are an important component of constant energy collection (Hansen & Butterfield, 1993, p. 135).

Physical structures aside, there are many other challenges wind energy suppliers must overcome to develop wind energy plants, including regulations and opposition groups, specifically in the United States. There are many groups in the US who claim they would prefer wind energy, to more traditional energy forms that they consider 'dirty'. Although as this case study illustrates, these environmental groups want wind energy, but just not in their backyard.

CAPE WIND—CASE STUDY

One example of the political, legal, and regulatory challenges facing wind power can be found on the Northeastern coast. Cape Wind, the proposed offshore wind development in the Nantucket Sound near Cape Cod, encapsulates how the political difficulties caused by green regulations and opposing environmental organizations can result in a standstill for wind projects in the United States. In terms of green development, both supporting and opposing organizations have worthy aspirations. Both the company proposing the project and groups opposing it declare themselves as bona fide green organizations. Local green organizations starkly oppose it, yet nationally recognized green organizations like the Sierra Club and Greenpeace have voiced their support for it. Both have reached out on social networking sites to try to get people engaged. The similarities continue, but the fact remains that their objectives are antithetical. To understand the political dilemma, an investigation of the circumstances surrounding the controversial proposed development is necessary.

The proposing company, Cape Wind LLC, is a joint venture between Energy Management Inc. (EMI) and Wind Management LLP. The project has been largely spearheaded by the efforts and funding of the president of Energy Management, Jim Gordon. Gordon gained an appreciation for the necessity of finding alternative energy sources after observing economic and societal disruptions caused by the oil embargoes in 1973 and 1974. He entered the energy business by founding EMI; a marketing company devoted to developing products for efficient energy companies, and began working on various energy conservation projects. The company adeptly used high demand for clean energy to find temporary success, but when prices fell, demand for their services fell as well. It became apparent that EMI would manage to survive only by drastically changing their business model (Burnett, 2009).

In 1978, Congress passed the National Energy Act, which provided various incentives for small energy producers to compete in the energy production market. EMI took advantage of the legislation by entering into the realm of energy production (Kimmell, 2011). Having gained market experience in the energy sector, EMI began innovating and hiring its own engineers. It entered the market in New Hampshire, building a biomass plant fueled by wood chips. It then built several natural gas powered plants, and became a pioneer of more efficient gas-fired electric generators.

The company was successful, both financially and in developing cleaner power. Its plants also provided hundreds of jobs. A total of seven plants were built in the New England area. Jim Gordon had made his fortune, and could have comfortably retired. Both Gordon and EMI had entrepreneurial fuel left in the tank, however, and in the decade before the turn of the century their focus turned away from efficient versions of traditional energy sources and began to emphasize alternative energy sources. The company invested in and explored a variety of different alternative power sources. Gordon realized that the Boston-based company was not surrounded by natural resources in the form of oil, gas, or coal, but was saturated by a viable form of alternative energy: coastal wind.

The company considered many locations for the wind farm, exploring the Northeastern United States for roughly a year. An ideal location needs to have more than just a steady supply of wind; it takes a significant amount of open space to place the number of turbines required to be economically viable. As the East Coast is densely populated, attempting to find enough land was difficult. Thus, the company's attentions turned away from the coast to finding an offshore development site. It was a bold decision: though offshore wind farms had been used in Europe since the early '90s, offshore developments in the United States had never before been completed.

In addition to having a constant wind supply, viable offshore farms would need to be close enough to allow transmitting power generated to the grid via underwater cables. To avoid public condemnation, it would need to be a couple miles off the coast; ecological analysis evaluating impact on wildlife would be necessary to avoid the criticism of environmental organizations. To receive the required approvals, Gordon knew that the selected location would need to be out of shipping lanes, ferry routes, and even flight paths. Despite seemingly endless miles of coastline, few locations seemed viable for accommodating the needs of an offshore wind farm (Burnett, 2009).

One location in particular seemed attractive to the company: Horseshoe Shoal, in the Nantucket Sound of Cape Cod; it seemed to meet many of the necessary requirements. It was far, but not too far from the coast, with the closest town, Mashpee, located 4.8 miles away. The shoal was particularly shallow, thus it was both largely removed from important shipping lanes (although noncommercial boats still use the area) and a viable candidate for use of "monopole" technology, a relatively cheap technique of supporting turbines that works best in shallower waters. Finally, it had an excellent wind supply, averaging just shy of 20 miles per hour. Capacity factor is a term used to describe how often wind speeds are optimal for a wind farm. The capacity factor of the Nantucket Sound is 37%, higher than the average for on-shore wind farms, which is at 34% (Kimmell, 2011, p. 201). Jim Gordon grew up in South Yarmouth, the town closest to the development, which gave him a background understanding of the area. As an extra bonus, the site, although surrounded on most sides by land, was at least three and a half miles away from each coast. The distance that meant the site (excluding the power cables and other infrastructure) would be located in federal waters, slightly reducing the potential political barrier of state jurisdiction (Burnett, 2009).

With a location in mind, it was easier for the company to calculate the logistics of the project because they could begin to account for local regulations. The proposed development covered twenty-four square miles, or about five percent of the total area in the Nantucket Sound. The traditional, horizontal axis turbines were to be located between four to eleven miles from the shoreline, and placed several thousand feet apart, a distance that amounts to roughly six to nine football fields of separation. The turbines were massive; the blades range from 75 to 440 feet off the water. For perspective, each forty-story turbine would extend over a hundred feet higher than the Statue of Liberty, pedestal, and all. One hundred and thirty turbines are planned for construction, thus it's important for the reader to realize the project was, and is, no small investment. Simply constructing the farm will take an estimated two years. Additionally, the estimated cost of the development is $2.5 billion, noticeably more than a comparable fossil fuel plant (Cape Wind: America's First Offshore Wind Farm on Nantucket Sound-Frequently Asked Questions, n.d.).

Its large size gave it not only a relatively large generating capacity, but would make it the largest off-shore wind farm in the world. For perspective, the would-be second largest off-shore farm, near the UK, was about 40% smaller (The largest offshore wind farm opens off Thanet in Kent, 2010). At peak generation, the Cape Wind turbines would generate 454 megawatts of alternative electricity, or enough to provide for the needs of 420,000 homes. On average, however, the expected amount of electricity to be generated from Cape Wind was about 170 megawatts, enough to provide for 160,000 homes, or 75% of the average electricity demand for Cape Cod, Martha's Vineyard, and Nantucket Island Combined (Cape Wind: America's First Offshore Wind Farm on Nantucket Sound-Frequently Asked Questions, n.d.).

With optimistic aspirations, Cape Wind began the lengthy permitting process, projecting that the development could be operational as early as 2004. Gordon assumed that the mostly liberal Cape Cod area would provide ample local support for an alternative energy effort like Cape Wind. Unfortunately for Gordon and Cape Wind, the proposal created an intense political firestorm, with attacks coming from both the left and the right and at both the local and national level. Very quickly it was realized that the project wasn't going to be an easy sell, if it was to be successful at all. Both sides dug in for a fiery legal battle that lasted nearly a decade.

The opposition groups were forming even before Cape Wind officially submitted the proposal for governmental review. The primary opposition group, the Alliance to Protect Nantucket Sound, formed in 2001 as a 501(c)(3) nonprofit organization dedicated to the opposition of the Cape Wind project, and has spent $15 million and ten years fighting the proposed project. The Alliance has a powerful membership, including prominent politicians and business magnates. Members include Walter Cronkite (Neal, 2009), former Massachusetts governor and two-time presidential candidate Mitt Romney, and Attorney General Tom Reilly. Dan Wolf, President and CEO of Cape Air (Cape Wind, 2004), as well as convenience store baron Christy Mihos, have pledged their support, while billionaire oil heir William Koch has donated at least $1.5 million to its cause (Doyle, 2006).

The membership of the Alliance has attracted some criticism. Many detractors of the alliance note that many significant supporters have property in the Cape Cod or Martha's Vineyard area, and that the Alliance to Protect Nantucket Sound is simply a front for them to fight against the construction of eyesore wind turbines in their scenic home and vacation areas- a classic case of NIMBYism. Though NIMBYism (they prefer to call it "preventing visual pollution") is perhaps a motivating factor of the actions of the Alliance, not all members and contributors are wealthy coast dwellers (Rosenthal, 2011). Certainly, ad hominem attacks can't dismiss all of the green issues that have resulted in political discord over the Cape Wind project. The conflict is better understood by examining the legal battles that occurred during the approval process of the Cape Wind development.

As of the first half of 2011, Cape Wind had been reviewed by seventeen different government agencies, they have spent $41 million (it's rumored that much of it came from Gordon's own pockets) and has encountered seven federal and four state lawsuits (Maroney, 2011). The uphill climb began in 2001, when Cape Wind applied for a permit under Section 10 of the Rivers and Harbors Act of 1899, and was placed under the environmental review of the National and Massachusetts Environmental Policy Acts. The US Army Corps of Engineers, who presented a favorable draft of an Environmental Impact Statement in a joint effort with the Massachusetts MEPA office, handled the application. The agencies allowed public comment periods, during which the draft was heavily disputed. Enemies of the project in Congress, such as Senators Ted Kennedy and John Warner (critics pointed out Kennedy either owned Cape Cod property or had friends and family who did) proposed barrier after barrier by raising the environmental standards the project had to meet and mandating often-redundant re-evaluations. The result: delaying the approval process by years.

All of the effort spent reviewing the project ended up being painfully meaningless. The Energy Policy Act of 2005 transferred regulatory authority of offshore energy projects to the Minerals Management Service, a bureaucracy in the Department of the Interior. This bureaucratic delay was a victory for opponents of Cape Wind. The Act was an attempt by Congress to establish a more orderly system for permitting offshore wind farming in the United States. It functioned, however, by removing permit authority from an organization that both had a favorable view of the project and had already made significant progress in the permitting process.

In the meantime, Cape Wind continued its quest by acquiring the necessary approvals at the local and state levels, which involved crossing a tremendous amount of red tape. Among others, the project needed a Chapter 91 license and water quality certification from the Massachusetts Department of Environmental Protection, access permits from the Massachusetts Highway Department for work along state highways, a license from the Executive Office of Transportation for a railway crossing, orders of conditions from the Yarmouth and Barnstable Conservation Commissions, road opening permits from Yarmouth and Barnstable, and finally, approval from the Cape Cod Commission.

The opposition applied significant amounts of pressure at the various stages of approval, even on the state level. The pipes that crossed through state waters needed approval by the Massachusetts Energy Facilities Siting Board. In May of 2005, the Board approved the application to build the wind farm. The process was so prolonged it took them three years, significantly longer than other project approval times. Most of time disparity occurred over questions regarding the environmental impact of the pipes, so despite the fact that the pipes had no obvious negative impacts on the environment, as demonstrated in multiple studies.

The gubernatorial election in 2006 proved to restore confidence in the Cape Wind project, which had been failing as a result of the strong political opposition. Deval Patrick, the candidate in support of the development, was elected, and he appointed Ian Bowles as Secretary of the Executive Office of Energy and Environmental Affairs. Bowles radically changed Massachusetts' permitting process, and Cape Wind gained the approval it needed. In 2009 the Massachusetts Siting Board even used its authority to override the objections of local agencies, finally completing the state and local approval process.

As the project's approval was coming together at the state and local levels, opponents turned their focus to preventing approval at the federal level. Another victory for opponents came when Cape Wind was forced to comply with the National Historic Preservation Act, which mandated that federal permitting agencies have to consider the effect of federal actions on properties listed on the National Historic Register. Although this had been completed under the environmental review process, an additional historic location was added for consideration. In the end, to occupy less space, the turbines had to be rearranged and the total number of turbines was dropped from the initial 170 to 130.

The conflict became more complicated in 2009 after the Mashpee Wampanoag Tribe petitioned that the entirety of Nantucket Sound should be on the National Historic Register as "traditional cultural property." They contended that, as the Tribes participated in "sunrise ceremonies," which involved the view of the Nantucket Sound from the East, the views should be preserved. Unfortunately for Cape Wind, the Keeper of the National Registry upheld the Tribe's opinion.

Neither the Wampanoag Tribe nor the State Historic Preservation Officer were willing to enter discussions or make compromises that would allow the completion of the Cape Wind development, arguing that Cape Wind should simply start over with another location. Finally, Secretary of the Interior Kenneth Salazar referred the issue to the National Historic Advisory Council for a recommendation on how the project should proceed. In April 2010, the Council recommended that the Secretary deny the project's approval. In a daring political move, the governors of Massachusetts, Rhode Island, New York, New Jersey, Maryland, and Delaware wrote Salazar, advising him to reject the advice of the Council. In response to the Council, they argued that if the decision was made on the grounds brought up by the State Historic Preservation Officer or the Wampanoag Tribe, it would become virtually impossible to build an off-shore wind farm anywhere on the East Coast.

The letter from the governors was quite influential. That same month, Secretary Salazar rejected the Council's recommendation, and issued a Record of Decision that granted Cape Wind supporters the victory they so desperately needed. Salazar's decision allowed for final permits to be issued by the end of that year, meaning that the Cape Wind offshore project is fully approved for construction (Kimmell, 2011, p. 225).

The story of Cape Wind is fascinating because of the perseverance of both the proponents and the opposition of the development. Both sides maintained vitriolic energy for an entire decade of legal struggles, over what a casual observer would perceive as a harmless alternative energy project. Though the project has been approved for construction, the Alliance to Protect Nantucket Sound is backing several different lawsuits challenging various components of the project; even suing utility companies for raising their rates to accommodate for the slight increase (about $1.50) per household that purchasing power from Cape Wind creates. Cape Wind only has a purchase agreement for 50% of its power, and the Alliance is fighting to prevent the rest of it from finding a purchaser (Environmental Management, n.d.).

Ironically, the Alliance is trying to generate grassroots opposition to the project by using "green" arguments. They assert in their informational leaflets about the project that the Cape Wind development harms birds and marine life and creates an expanded risk of an oil spill (the project requires a ten-story platform with 40,000 gallons of oil for operations). They assert that Jim Gordon and Cape Wind aren't interested in greening the earth as much as padding their wallets through higher energy rates (Cape Wind Threats: The Environment, n.d.).

Cape Wind also claims, however, to be the environmentally correct party. They claim that using wind energy instead of gas-fueled generators will reduce the number of oil spills (two have already occurred in the area as a result of tankers running aground while shipping fuel to power plants). In response to the danger to marine life, they argue that the greatest danger to marine life is climate change, and wind power cuts down on carbon emissions that contribute towards it. They view the Alliance as a group of wealthy landowners who may or may not want wind energy, just not in their backyard. Who is "correct" might be arbitrary; a clear example of green versus green.

ANALYSIS

The Cape Wind story is an example of how provincialism can derail a alternative energy project. It is a story of a case that has received local condemnation, national attention, and support, and provides a global public good. Local individuals and groups in this story exemplify the elementary use of environmental institutions to protect their interests: eleven lawsuits in total. These interests do not necessarily need to be personal gains in the form of wealth, or a 'big win' for conservation groups by preventing a new energy plant from being sited. But these wins can be as small as limiting the size of a alternative energy plant, or in this case limiting the acceptable placement area and completely removing the final number wind turbines from the plan.

POLICY ARENA

The disputed Cape Wind Project is governed by a series of national institutions, Massachusetts' institutions, and local institutions, all of which do not necessarily coincide. Federal laws designed to find a balance between man and nature are often at odds with greater policy goals. For example, the Department of Energy's goal of having 20% of the nation's energy come from wind energy by the year 2030 conflicts with local protests over this wind farm (Energy Basics: Wind Power Animation, 2008). Due to the legal constraints of federalism, any state environmental laws must be more stringent than those at the national level; otherwise the national regulation trumps. Local environmental regulations may also constrain how materials are gathered, constructed, and used, always with more limitations than the previous two levels, although state governments can overrule local statutes.

The area chosen for this site was not selected randomly. The favorable wind conditions appeared to make it a worthy investment, meriting the time, money, and tedious regulatory process. Horseshoe Shoal is close enough to development to make it worth the transmission costs. It is far enough away so a not to be a real eyesore, it does not block shipping routes, and the ocean floor is shallow enough to properly place wind turbines. Additionally the wind capacity is higher than the coastal average, making the energy collection possibilities immensely satisfactory.

The Cape Wind developers understood these complications and made a plan designed to comply with the written rules, at all levels, simultaneously. Unfortunately, there are other environmental institutions against which companies cannot entirely prepare themselves, the legal system and informal institutions built in local communities. While those at federal and state agencies primarily make land use-decisions, lawsuits have become the informal tool of small groups and individuals. Thus, decisions over whether or not alternative energy plants will be sited are never entirely in the hands of federal regulators. Additionally, companies siting wind energy may not always have a complete understanding of the formal and informal institutions where they are developing, as they are subject to constant change.

KEY STAKEHOLDERS

Local residents have a vested interest in preventing Cape Wind's siting because it could affect property values, or at least personal valuations of the area. The Alliance to Protect Nantucket Sound, and other similar groups, tried and will continue to try to sway public opinion by arguing for the safety of the environment or its inhabitants. It cannot be ignored, however, that these groups stand to benefit if this siting is prevented. Despite the fact that these wind turbines would be far enough from the coast that they would barely be noticeable on the horizon, families that have owned beachfront property from years see this wind farm as a threat to their property.

The Alliance to Protect Nantucket Sounds obviously values free access to this area, both physical and visual, as demonstrated by the $15 million they have spent thus far. It is also interesting to note the duplicity of some politicians like Senators Ted Kennedy, John Warner and Mitt Romney, all politicians from Massachusetts who own land, or have family with land, near the proposed site. On the national stage they have each called for an increased use of all alternative energy, including wind power. But when it comes to siting wind energy in their 'back yards', suddenly they are on the defense. It is not hard to imagine that other local citizens have also called for an increased use of wind energy, but they demonstrate strong opposition to this project because they feel it would negatively impact them.

Supporting this project, quite obviously, is Cape Wind LLC. Instead of focusing on the company, however, more can be learned from focusing on Jim Gordon. While Gordon has an encouraging story describing why he got involved in alternative energy, as well as a long history involving his promotion of biomass and more efficient gas systems, it is important to remember that he gains from this project succeeding. Not only does he gain happiness from finally seeing his dream put into reality, but Gordon is also rumored to have invested much of his own money in getting this site through the legal system. Unfortunately he suffered from provincial naïveté and did not anticipate the blowback his plan received.

Five years into this regulatory review Cape Wind received some much-needed help from a couple of state politicians. Deval Patrick's election as Governor gave Cape Wind the advocate they needed at the state level. Ian Bowles, put into place at the head of Massachusetts' main energy regulatory board, would then be able to help with the state regulatory institution, the place where local groups were easily able to pose stumbling blocks for the project previous to Bowles appointment.

Since there is little chance in this development of seriously harming flora or fauna, any groups or individuals who do not own property in the Cape Cod area have no reason to oppose the measure. In fact, endorsement of this project will often be beneficial for politicians or any groups hoping to appear 'green' and appeal to a certain type of voter. Even national conservation groups that often oppose siting new energy, when they feel the siting will cause environmental damage, such as the National Sierra Club and Greenpeace, support this wind farm. Also, these groups are not likely to suffer any losses or be harmed by this project, so there is no disincentive to support it. State leaders supported the plan, because the overall effects for the state would have been good. While a small group of constituents opposed it, this wind farm could mean clean power for many more citizens, and clean air for all those in the state.

RULES IN USE

In order for the Alliance to achieve their objective of preventing these windmill 'eyesores' from cluttering their coastal views they have to use the institutions available to them: environmental laws. Even when groups opposing the siting of an energy plant know they do not have a very solid case, it is still in their interest to file lawsuits. Filing suits affords them several options. First, environmental groups could win and completely prevent the siting of a alternative energy site. Naturally this is their preferred outcome. Second, a lawsuit can force the energy company or agency to make adjustments; these adjustments can range from small construction elements to complete overhauls of their plan. Such was the case when Cape Wind was forced to do with their farm after suits were filed under the National Historic Preservation Act. Third, companies can go bankrupt while attempting to site a new plant. Most companies do not have enough money to face years of legal battles and fees and forced so far into the red that construction doesn't even begin. The final option is case dismissal; cases brought against alternative energy companies or agencies are completely dismissed because they lack merit. Currently Cape Wind enjoys national and state support, enough so that it is unlikely the Mashpee Wampanoag Tribe's complaints will cause the site to be dismissed completely, however, it is possible that Cape Wind's plan will have further stipulations placed on it.

As soon as Cape Wind applied for their first permit under the Rivers and Harbors Act and comments were requested on Cape Wind's EIS, the outrage became apparent. Under the Rivers and Harbors Act Cape Wind was to work with the US Coast Guard to ensure that they weren't constructing any bridge, dam, or similar structure across navigable waters. Cape Wind also had to ensure that they were in compliance with the National Migratory Bird Treaty Act and that their wind turbines would not disturb any birds traveling between Canada and the US.

Cape Wind would also have to go through the NEPA process, demonstrating that their actions would not have any damaging effects on the environment. Due to all the controversy surrounding Cape Wind, Alliance members would have ensured that Cape Wind was forced to go through the entire process. Critics took the US Army Corp of Engineers' EIS to task (part of NEPA), complaining that the review had not been thorough enough, Senators Kennedy and Warner intervened, using their political connections to ensure that several re-evaluations were required and the process was delayed. After the NEPA processes were completed, the Mashpee Wampanoag filed their claim with the National Historic Registry. Since the Wampanoag Tribe had stated their unwillingness to bargain it was set to be a tough fight. Cape Wind however, did have the federal government's support, giving them the needed authority to move past this roadblock.

In addition to these federal acts, Massachusetts has many of its own strict environmental regulations. Gordon attempted to circumvent many of these by only placing turbines in federal waters; however the state still had many acts that the Alliance could use to complicate his efforts. Cape Wind also had to perform the NEPA process, only under Massachusetts' tighter regulations. They had to seek approval under the Energy Policy Act's guidelines from the state Environmental Affairs Office. Approval also had to be granted by the Massachusetts Energy Facilities Siting Board for any pipes running across stateliness. Local detractors attempted to block Cape Wind's advancement through the state process in any way possible.

INFORMAL RULES

Not only did Cape Wind have to deal with a constantly changing federal regulation system, but a local system. Local groups were able to put pressure on state regulators, one tactic that was not as easily accomplished at the federal level. The Massachusetts Energy Facility Board intentionally prolonged the time Cape Wind was forced to wait for a permit, demonstrating that it is possible for local and state regulators to prolong the time for permits, in order to appease local crusaders. It wasn't until the installation of Ian Bowles that Cape Wind was able override the pressures put on local regulatory authorities and obtain the necessary local permits.

Not only does Cape Wind face extreme opposition from average, local citizens, the company also faces hostility from citizens who have the power to change the environmental institutions that regulated them. Over the past ten years environmental standards were raised at both the state and national level forcing Cape Wind to update their plans in order to comply. Additionally, the Energy Policy Act was passed in 2005; completely removing one of the agencies Cape Wind had dealt with for five years. Although this change was intended to create a more orderly system, it didn't grandfather in agencies or companies who were already in the process of permitting under the previous regime. Cape Wind was then forced to restart the process and undergo a longer permitting process than originally planned.

The Alliance to Protect Nantucket Sound was blessed with many well-connected and wealthy patrons. They were able to use these advantages to gain an edge. Asking for large donations the Alliance has waged a large and generally effective campaign against the Cape Wind project. Both sides have reached out and looked for support on the internet. These two groups have created massive web sites in addition to using social media to try to garner attention for their side of the issue. The earlier referenced video put together by the Daily Show is just one example of the ways in which this issue has been presented using non-traditional forms of media.

OUTCOMES

Before Cape Wind was able to gain favorable attention from Secretary Salazar, Deval Patrick and other state and federal leaders, the project seemed doomed. There were, however, several possible outcomes. First, the Wampanoag Tribe could have stopped the project if their appeal to have the area listed on the National Historic Register would have been approved. Also, new politicians could have come into office that didn't agree with the project and it would have stayed locked in legal battles, or much of East Coast could have been listed on the National Historic Register. As this project went through the regulatory process the process could have limited the proposal to such an extent that it was no longer an economically feasible project and it would have been abandoned. The Cape Wind company could have been locked in lawsuits for such a long time that it lost investors and went bankrupt, a fairly common occurrence.

The Cape Wind story demonstrates that energy companies or those hoping to site new energy plants can never take for granted how the local community will react. It seems logical that a generally liberal area such as Nantucket Sound would be more than willing to allow a wind farm right on its border. Jim Gordon and others had to learn that even the most liberal of citizens do not want new energy plants, alternative or not, positioned in their backyard.

Even though national policy makers set lofty goals for using wind energy, there are few local communities pushing for the installation of wind turbines near their homes. But, if this case study is representative of how most citizens feel about wind farms, that goal will be difficult to achieve. In this case the wind energy company was lucky to have intervention at the state and national level. If there wasn't some intervention at the national level, it is likely this area could be locked in controversy for another ten years. While that may give hope to other small companies fighting for the right to site a new energy plant, it is not reassuring on the macro level.

CONCLUSIONS

More than fifteen years ago, Robert Bradley Jr. wrote a piece for the CATO Institute titled, "Renewable Energy: Not Cheap, Not 'Green.'" The article records several problems that wind will face moving forward. Although somewhat dated, Bradley's concerns for wind are still relevant today given the state of the economy and uncertainty regarding future production tax credits and subsidies. Bradley wrote, "With. . .an impending electrical industry restructuring that could force all generation resources to compete on a marginal cost basis, wind power is a problematic choice for future electricity generation without a new round of government subsidies and preferences" (Bradley, 1997).

The implications of our analysis do not hold much promise for the wind energy industry. Even communities where a preference for wind power would be higher than other communities, wind farms are strongly opposed. Technical issues aside, federal, state, and local regulations make it difficult for even the most wealthy and powerful of alternative energy industry giants to site a new plant. Developers, landowners, and policy makers have many concerns to address if they want to move forward with wind development.

5   Solar Energy

INTRODUCTION

Luz International Ltd. built and operated nine solar-electric generating stations in the Mojave Desert during the 1980s. Those plants continue to operate but Luz International had to declare bankruptcy in 1991 as their state property tax exemption expired and other state and federal support programs ended. Two decades later, under many of the same directors including the founder Arnold Goldman, BrightSource Energy is building the $2.2 billion Ivanpah Solar Electricity Generating Station (SEGS) in the Mojave Desert. This plant is making headlines as "the world's largest concentrat[ed] solar power plant under construction" (BrightSource Energy News Releases, 2010).

One of Bright Source's biggest hurdles has been the desert tortoise, which is listed as endangered under the ESA. BrightSource officials knew the site they chose was desert tortoise habitat and worked out mitigation plans including installing fifty miles of fencing at a cost of $50,000 per mile. They obtained permission from the U.S. Fish and Wildlife Service to move up to thirty-eight adult tortoises off the site. Their permits allowed them to kill three tortoises accidentally each year during the construction phase of the project. It turns out that the original survey of tortoises underestimated the number of tortoises actually living on the site and in early 2011 state and federal agencies required that construction be delayed until a new environmental assessment was completed.

HISTORY

Solar power is the collection of the sun's light and its conversion into energy. Humans have been indirectly harnessing the power of the sun since the shift to an agriculturally based society. The Greeks, Romans, and Chinese first used solar power to light torches with glass and mirrors for religious purpose. By 600 AD, the science behind solar power had evolved enough to be used to heat homes and buildings.

French-Swiss scientist Horace de Saussure saw the potential to use the sun's energy in the early 1700's, and tested this potential through experiments. He built a small "hot box," or greenhouse, and measured temperatures inside the box. After rotating the box for several hours, he discovered that temperatures reached up to 189 degrees Fahrenheit. He revised the design of his box by adding several layers of glass and insulation to limit heat from escaping the box (Butti & Perlin, 2005). Following Saussure's experiments, researchers looked for new ways to improve the technology. Edmond Becquerel, for example, generated a continuous current using sunlight in 1839 (Fraas & Partain, 2010). He created the first known copper-cuprous oxide thin-film solar cell after submerging two brass plates in liquid and allowing sunlight to heat them (Fraas & Partain, 2010).

Solar energy technology took a leap forward with the discovery of selenium in 1873. Selenium is a solution that produces energy when it is exposed to light, making it an obvious choice when trying to find useful materials with which to convert sunlight into energy. Ten years following the discovery of selenium, American C.E. Fritts used it to develop the first solar cells, also known as wafers. Wafers continue to be used in today's solar cells and panels (The History of Solar, n.d.). Early pioneers of quantum mechanics, including Albert Einstein, developed the theoretical underpinnings of converting solar energy to electricity. In the early 1900's, Einstein published a paper on the photoelectric effect (work for which he later won a Nobel Prize), leading to further experiments and development of "photovoltaic technology". This photovoltaic technology made it possible to mechanically run equipment normally powered by electricity. In 1954, almost seventy-five years after Fritts' original discovery, Bell Labs created the first silicon solar cell capable of running everyday electrical equipment. Not only was Bell Labs able to duplicate Fritts' cell, but it also increased efficiency from 1% to 6% (Fraas & Partain, 2010). Research during this time also led to the design of solar-heated buildings. With the advent of the bipolar transistor, photovoltaic (PV) efficiency more than doubled to 15% (Green et al., 1999). This technology was used by NASA to power satellites and observatories, providing data to scientists about the earth's atmosphere (The History of Solar, n.d.).

Passivated Emitter Solar Cells (PESCs) were the next breakthrough. They were simple to make and remained the industry standard for almost ten years. They were developed out of efforts to 'passivate' or prevent electrons from escaping from the backside of the cell. These cells were 'passivated' across the entire cell, leading to over 22% efficiency (Green et al., 1999). By 1999 the PESC model was surpassed by the Passivated Emitter Rear Locally-Diffused (PERL) cell, which contains a thermally oxidized and newly texturized surface on the backside (Green et al., 1999). Another fairly recent change has been the use of small, inverted pyramids on the cell to capture more light. The application of miniscule layers of silicon on the cell has also increased energy output. Additionally, scientists combine layers of the photovoltaic material into one solar cell, creating higher energy efficient cells.

SOLAR ENERGY BASICS

The Photovoltaic Effect, discovered in the early 1900's, uses cells which function as semiconductors. It is a basic process through which photovoltaic (PV) cells convert sunlight into energy and electricity. Edmund Becquerel, a French physicist, discovered the Photovoltaic Effect in 1839; he found that certain materials would produce an electric current when submerging two brass plates in liquid and exposing them to light (The Photovoltaic Effect, n.d.). When these cells are exposed to sunlight, energy is reflected, passed through, or absorbed. Photovoltaic cells convert the absorbed energy into usable energy. These Photovoltaic cells are typically used in solar panels seen on roofs (Energy Basics: Photovoltaic Cells, 2011).

In order to create effective technologies and mechanisms, a variety of cellular materials are used, depending on the purpose of the solar technology. These elements are used in both pure forms and compound combinations to effectively convert the light into energy. Silicon, an element used in both early solar devices and today, was discovered in 1824 by Swedish chemist Jöns Jacob Berzelius (It's elemental, n.d.). Silicon is found naturally, or can be produced by heating sand with carbon at 2200 degrees Celsius (It's elemental, n.d). Semi-crystallin silicon is an alternative to silicon. Although it is not as widely used as silicon because the flow of electricity is not as effective, it costs far less than using silicon (Energy Basics: Photovoltaic Cells, 2011). Each type of silicon is produced differently based on its intended use and solar cell technology. The primary types are described below:

Single-Crystal Silicon is the highest-purity silicon, created by melting the single crystals and then cooling them slowly. This process allows for the growth of a new rod or "boule" as the material solidifies slowly. Three processes can grow a "boule": the Czochralski, Float-Zone, and Ribbon silicon processes. The Czochralski process dips a crystal into molten silicon and then slowly pulls it out, creating crystals. The crystallization of the "boule" allows photovoltaic charges to flow and act as conductors. In contrast, Float- Zone silicon crystals are put through an electromagnetic coil, creating higher purity crystals than Czochralski crystals. These crystal rods must then be sliced into thin wafers once single crystal rods have been produced. Unfortunately, this slicing process wastes 20% of the silicon. Both the Czochralski and Float-Zone methods are expensive and complex, but well developed (Energy Basics: Types of Silicon used in Photovoltaics, 2011). Ribbon silicon is a single-crystal silicon technology which forms into thin wafers and is less costly to produce than other silicone forms. The tradeoff is that the efficiency of Ribbon silicon is approximately 16.2% (Alternative Energy, 2010).

Multicrystalline silicon is less expensive than single-crystal silicon, but far less efficient due to the lower grade material used. Because of its square molecular shape and structure, it is able to fit compactly into a photovoltaic module. These variations of silicon are based on the purity of crystals, molecular structure, and its ability to absorb solar radiation.

Amorphous silicon is a photovoltaic compound commonly used in calculators and wristwatches, and can be produced in lower temperatures. It's economically advantageous because the production cost is significantly less than it is for photovoltaic. Amorphous cells are also thicker allowing them to absorb 90% of the sunlight they are exposed to. Amorphous cells are not stable, however, as output decreases by as much as 20% over time (Energy Basics: Photovoltaic Cell, 2011).

Today there many innovative ways of building photovoltaic cells. One example is Polycrystalline Thin Film, which is a roll-based laminating system producing amorphous silicon solar cells. These cells are thin cells, which have an advantage over thicker and chunkier cells because they require less material and are generally lighter and more flexible than thick cells. The downside is that they are less efficient than thicker cells. Another example is Printable Solar Cells. These cells are produced from ink, creating semiconductors or solar cells. By using nanostructure materials, high quality electric films can be printed in less time and lower cost than manufacturing thin-film solar cells (Nanosolar, 2011). Once it has been printed on prepared aluminum foil, the cells are sorted based on function, then assembled and connected into panels, laminated, and covered by glass.

An innovation to solar energy is polymer solar cell technology. It is is light and flexible, making it possible to gather light through ridges. Though the cell has good electrical transport, the overall performance has been poor because of small charge and short circuit. It does however, have a high conversion and efficiency rate, with the ability to trap 20% more light than flat solar cells. Plastic solar cells are also popular because of their durability and lightweight features. Because of its flexibility, these plastic cells can be placed around surfaces and used for roof tiling and siding. Plastic cells have been used on vehicles and other machinery as well. Solar concentrators that use dye are also used to absorb and trap more light to transport it for conversion in lower layers of the solar cells (Kroon, 2009).

ADVANTAGES AND DISADVANTAGES OF SOLAR ENERGY

The price of photovoltaic solar energy systems, in comparison to coal, is relatively high, exceeded only by offshore wind and solar thermal systems. There are, however, a number of incentive programs including tax credits, rebates, low-interest loans and grants provided by government programs to encourage the use of solar energy (Types of Solar Power, 2012).

According to the U.S. Department of Energy, alternative energy like solar energy will not be competitive until "soft costs" can be reduced or diminished. Soft costs are the expenses associated with permitting, financing, interconnection, and inspection. They are estimated to create 40% to 50% of the overall cost with use and installation. In addition, there is no consistency between local, state, and national government regulations (DOE Awards $12 Million, 2011).

President Obama's "race to the top" campaign has made the goal to make it easier to obtain required solar energy siting permits, as well as create a number of economic incentives to increase solar energy projects. Several grants were awarded throughout the United States in 2011 to financially assist in various projects conducted by local government and jurisdiction in an effort to reach nearly 51 million Americans in twenty-two locations. This campaign's efforts are to make solar energy more competitive with other sources by 2020 (Simmons, 2011). One such project, the Department of Energy Rooftop Solar Challenge, encourages local governments, utilities and installers to compete by implementing plans to reduce administrative barriers that restrict solar energy devices installations and to help increase the global market by making solar costs competitive (Simmons, 2011).

IVANPAH POWER-CASE STUDY

Arnold Goldman, founder of BrightSource Energy Inc., is a man with a vision of a world powered by natural sunlight. He remembers waking "up one night feeling miserable and [coming] to the conclusion that if I had to work most of my life, at least I wanted it to be valuable" (Meet Brightsource's Arnold Goldman, 2010). What could this young man do to make his life valuable? Through alternative energy, specifically, solar energy.

Goldman graduated from the University of California-Los Angeles with a bachelor's degree in engineering and minors in philosophy and economics, after which he earned a master's Degree in computer science at the University of Southern California. After spending five years in computer development with the military and working for defense contractor Litton Industries, Goldman, along with a group of scientists from the California Institute of Technology and Xerox, founded the first word processing company in the United States in 1972: Lexitron. In 1979, the entrepreneur then founded Luz International, Ltd. in Israel, which later expanded to California (Meet Brightsource's Arnold Goldman, 2010).

As founder and CEO of Luz International, Ltd., Goldman headed the largest solar power plant of its kind in the 1980s, engineering and building nine large solar power plants in the Mojave Desert of California. This system of plants is known as Solar Energy Generating Systems (SEGS) and produces a combined output of 354 megawatts of electricity, enough electricity to power 140,000 homes (Ivanpah, 2012).

The 1973 oil crisis sparked an increase in government support for alternative energy: OPEC (Organization of Arab Petroleum Exporting Countries) had sought to display their disapproval of the United States' decision to re-supply the Israeli military during the Yom Kippur war by placing an oil embargo on the U.S. Four years later, President Jimmy Carter urged the nation to support alternative energy efforts to protect against future incidents like the 1970s oil crisis. He specifically listed fuel conservation, the development of alternative fuels, and a reduction in dependence on foreign oil as prime goals (Carter, 1977).

One of the Carter Administration's main goals was to "use solar energy in more than two and one-half million houses," by 1985 (Carter, 1977). Just two years after Carter's speech, the Iranian Revolution shattered what little stability Iran had and gave way to a new government. The new regime exported oil inconsistently and at a significantly lower capacity, sparking the 1979 energy crisis. The United States' oil supply was again severely impacted as a result, reinforcing Carter's message that alternative energy was a brighter and safer option than the current dependence on unreliable foreign oil.

The favorable politics and the relative novelty of the ESA allowed SEGS to be built with little opposition. Opposition facing green energy companies was considerably less concentrated and established during Luz's peak than now. Luz Industries began constructing their solar plant during the perfect window of opportunity, sandwiched between a high public approval, as a result of the energy scare in the 1970s, and little organized resistance.

By the late 1980s, government policies promoting and enabling the construction of solar power plants began to expire. These policies included federal and state tax credits (some as high as 10%), exemption from property taxes, and special depreciation schedules. While continual construction of plant after solar plant gave the impression that Luz International Ltd. was still a lucrative business, the company was barely treading water, selling one plant to raise money for the next plant and relying heavily on tax credit renewals.

By the late 1980s, state and federal solar energy tax credits and investment tax credits were beginning to expire, and it seemed unlikely that politicians would be willing to extend them. As a result, Luz began building plants in increasingly shorter time periods to qualify for the tax credit: SEGS IX was built in 7 ½ months in 1990 to fit the restrictive time stipulations of the Federal tax credit. Luz International had to pay huge cost overruns in order to rush construction of SEGS IX, which was the final blow to the company. Finally, in late November of 1991, the company filed for bankruptcy, halting construction of a tenth plant and permanently dissolving plans for an eleventh and twelfth (Berger, 1997, pp. 38–45). After selling SEGS IX to NextEra Energy Resources, which already owned SEGS III–VIII, Luz International completely disbanded.

In 2004, Arnold J. Goldman reassembled the original team of Luz International Ltd. to form Luz II Ltd., later renamed BrightSource Energy, Inc. After the relatively successful SEGS I-IX solar plant, which continues to produce seventy percent of all solar power currently produced in the United States, BrightSource proposed several new projects and by early 2012 had several developments underway, including the Ivanpah Solar Electricity Generating Station (ISEGS) in the Mojave, which, if finished, will be the largest solar plant built since the Luz International solar power plants in the 1980s.

The Ivanpah Solar Electricity Generating Station consists of three separate solar power thermal plants, which have a planned capacity of 392 MW. It has received a $1.6 billion loan guarantee from the U.S. Department of Energy and has been singled out by the President as a "revolutionary. . .state of the art facility," that will create jobs and help the nation achieve its clean energy goals (Ivanpah, 2012). The project will consist of 4000 acres of mirror fields, fences roads, fences, and transmission lines. An acre is roughly equivalent to a football field without the end zones. So, the Ivanpah facility will be about the size of 4000 football fields. The mirror field will contain nearly 200,000 glass mirrors, each the size of a garage door. Those mirrors will focus on three 460-foot-tall "power towers." The project is located a few miles from the Mojave National Preserve, the nation's third largest unit of the National Park system outside of Alaska.

A great deal of opposition to the Ivanpah project and other potential solar projects on public lands has emerged. Kim Delfino, Defenders of Wildlife's California program director, cautions, "California is starting to see a new kind of gold rush, but this time it's going to be our wind, sunlight, and public lands that are up for grabs" (Navarro, 2009). A sampling of opposition from the green community can be found on an online blog, titled the "Mojave Desert Blog." Writers on the blog have many concerns, the first of which is the possible destruction of this diverse habitat. They also discuss their concern over potential harm to the desert tortoise and other native, rare plants in the area. These writers were also concerned that one solar plant would open the floodgates and soon the whole area would be developed. Finally, there were concerns raised over the possible loss of viewsheds and the "fragmentation" of the desert landscape (Mojave Desert Blog, 2010).

BrightSource chose an area of the Mojave Desert that is rich with ecological resources (Seltenrich, 2011). Besides the desert tortoise, there are at least twelve species of rare plants on the site (Clarke, 2009). Individuals such as writers for the Mojave Desert Blog writer and groups such as the San Francisco's "Desert Survivors," a self-described "affiliation of desert lovers committed to experiencing, sharing and protecting desert wilderness wherever we find it," make it their mission to ensure that such environmental impacts are correctly and fully reviewed (About Us, 2009). Members of these groups are passionate and committed to protecting their local resources from development of any kind.

While opposition from grassroots organizations is significant, the first major hurdle BrightSource faced was a strenuous review from the California Energy Commission. Their review process mandated under the California Environmental Quality Act was combined with the BLM's NEPA process in an attempt to expedite the process and increase efficiency. There were so many environmental concerns, however, that the process took three years.

BrightSource knew they were attempting to build in desert tortoise habitat and performed the preliminary biological survey necessary to move forward. That survey found just sixteen tortoises and, based on U.S. Fish and Wildlife Service formulas, BrightSource was granted permission to move thirty-eight tortoises and kill three a year (accidentally) during its three-year construction phase (Danelski, 2011). After three years, the California Energy Commission granted unanimous approval to the project and BrightSource agreed to the stringent environmental restrictions that included fifty miles of turtle fences designed to keep turtles out of the area, reducing the number of "power towers" from seven to three, and reducing the project's footprint by 12%. On October 7, 2010, Ivanpah received the BLM's Record of Decision (ROD), giving them the green light to start construction. Twenty days later, a celebratory commemoration marked the groundbreaking of the BrightSource Energy Ivanpah plant.

BrightSource employed as many as 100 biologists to make sure they complied with environmental restrictions; as part of their work, these biologists tried to determine a baseline tortoise population. By February, 2011, project biologists indicated that they had found many more tortoises than expected from the initial survey. Because of the survey, the BLM ordered the company to stop construction on the plant and authorized a new biological assessment to be performed. Results of the revamped evaluation were stunning: ninety percent of all non-adult tortoises would most likely be exterminated in the process of building the three power plants. Biologists estimated that would equal as many as five hundred forty seven individual underdeveloped tortoises per year. The rest of the tortoises—the lucky ones—including one hundred sixty resident adults, would be manually transferred to off-site pens to be held during the winter and later relocated to a permanent protected reservation. In total, the Ivanpah complex would destroy approximately three thousand forty four acres of desert tortoise habitat and displace over three thousand desert tortoises, killing many of those tortoises in the process (Seltenrich, 2011).

The U.S. Fish and Wildlife Service (FWS) ended up re-negotiating stipulations with BrightSource Energy in June of 2011 that would permit building plans to continue, if certain conditions were met. The FWS released a "new Biological Opinion. . .[which] includes new stipulations for translocating tortoise[s], as well as new requirements for protecting them from predators and increased monitoring and fencing. The FWS also required BrightSource to purchase more land as compensation for loss of desert tortoise habitat" (Service Issues Biological Opinion for Ivanpah Solar Electric Project; BLM Lifts Suspension of Activities Order, 2011). The first tortoise relocation at Ivanpah occurred on October 10, 2011. The company posted a picture of the "healthy female tortoise" on their blog post after the successful translocation, describing the process in detail (First Tortoise Translocation at Ivanphah, 2011).

Construction of the solar power plant continues, as project biologists work to ensure the safety of each tortoise found on the site, following the guidelines illustrated by the FWS. In a blog post, BrightSource outlines the steps taken in each tortoise's "individual 'disposition' plan, which tracks the tortoise's health, activity, [and] habits." Afterwards, tortoises undergo an all-inclusive medical assessment to determine whether the animals are healthy or if they may "carry a potentially-fatal respiratory disease that is prevalent and contagious among the desert tortoise population." The area chosen as the relocation land for the tortoises is reviewed multiple times previous to actual transfer to evaluate "if the "recipient population" . . . can safely accommodate additional tortoises." The animals are then only translocated during specific intervals in the spring and fall and when temperatures are between sixty-five and ninety-five degrees Fahrenheit (First tortoise translocation at Ivanpah, 2011).

While some opposition continues against ISEGS, BrightSource continues their plans with backing from government and alternative energy proponents worldwide and applause from environmentalists for managing to produce alternative solar energy while also saving the endangered desert tortoise. The plant is set to finish construction in 2013 as the "largest solar plant under construction in the world" today (First Tortoise Translocation at Ivanpah, 2011).

ANALYSIS

The BrightSource Energy case study shares much in common with the Cape Wind narrative. Both stories feature an idealistic planner, hoping to implement their big dream of an unprecedented alternative energy plant. Both suffered from the same providential naïveté, expecting their communities to be just as excited about their project as they were. In addition, both men in charge of the project lived closed to the areas where they are attempting to site alternative energy plant, and thought they knew the area well. This case story, however, occurs on the other side of the country, and deals more heavily with endangered species. Additionally, the BrightSource Energy case demonstrates how many environmental institutions and those promoting them have changed in the last thirty years.

ACTION ARENA

Similar to the Cape Wind proposal, the Solar Electricity Generating Station is also governed by a multi-tiered system of environmental institutions. California has its own set of environmental institutions that have proliferated over the past several years, similar to those in Massachusetts. When BrightSource went about re-applying for permits and gaining permission to site in California in 2010, the process was very different than it had been twenty years prior. Many new environmental laws had been passed, as well as the interpretations and reach of previous laws had been changed in court rulings. Additionally, the way that regulators fulfilled these laws had changed, as have regulators.

Not only was the permitting process different, but the community was as well. When BrightSource Energy, then Luz International, originally sited their Solar Electricity Generating Stations in the 1980's there was little resistance. After the high oil prices of the 1970's community members were happy to see alternative energies being sited. But since then so much has changed. Conservation groups have spread, and they have gotten good at using litigation. Additionally most communities now have a sour taste in their mouth when it comes to siting huge energy developments. While there is still much rhetoric from politicians on the national scale regarding the need for alternative energy, most citizens do not feel the pressure they once did to race towards new energy establishments.

One way that politicians demonstrate their desire for more alternative energy is through some combination of subsidies and alternative energy tax credits. For many companies this is what makes producing alternative energy worth the effort and time. Luz International survived for many years only because of these credits, but was unable to continue once they had ended. A similar situation might not be too far away. Next year many tax credits come up for alternative. If the recent budget negotiations are any indicator, it is likely there will be a hard-fought battle over whether to renew these, and how to go about renewing them.

KEY STAKEHOLDERS

As mentioned above, Arnold J. Goldman is yet another man with green dreams. His is a vision of the world's largest solar plant, and a chance to redeem his earlier solar projects. Similar to Gordon (Cape Wind CEO), Goldman is blessed with years of experience in the field. He understands how to petition politicians, work with bureaucrats, and promote his dream to community members and investors. Aside from getting the opportunity to finally achieve his dream, Goldman would obviously profit from this, financially. Goldman also has the chance now to demonstrate his improved solar energy collection system. His proposed plant in the Mojave Desert would do just that.

While Goldman has been working on the engineering behind his projects, the opposition has had time to form. Many groups have joined forces and used the internet, specifically their Mojave Desert Blog, to tout their concerns. They claim that while the Ivanpah plant will provide public goods, the overall operation will have a negative effect. These groups choose to focus on the effects that this plant would have across the entire desert, and maintain that this project will lead to a loss of viewsheds and natural landscapes. These groups also cite the potential damage this project could have on the desert tortoise and several critical plant species.

One supporter that Goldman and BrightSource Energy can begin to count on again is the government. Despite recent controversies, the Obama Administration has promised much in the way of subsidies and tax credits to those installing solar energy plants. Obama has even praised the Ivanpah plant, specifically. After the Solyndra embarrassment in early 2012, the Obama Administration would benefit from having a profitable solar energy plant established with the help federal funds his Administration helped to allocate.

RULES IN USE

At the state level, BrightSource Energy has faced much scrutiny from the California Energy Commission. One kindness BrightSource Energy can be grateful for was the decision to combine three processes from two separate committees into one. This sped up the process for BrightSource Energy, but also limited the opportunities for conservation groups to slow down or halt action on the siting. It has given these green groups less time to garner support during commenting periods. This seems to signal willingness on the part of the state and federal agencies to move this project forward.

Concerns over damage to desert tortoises and their habitat have also delayed the approval of the Ivanpah plant. The discovery of a larger number of tortoises than originally anticipated has caused difficulty for Bright-Source Energy and increased mitigating measures and renegotiations. It is a very possible scenario for the company to face a lawsuit charging them with Endangered Species Act violations.

INFORMAL RULES

As BrightSource Energy has discovered, the unspoken rules that govern energy siting have changed significantly in the past thirty years. Even though the country still faces high oil prices, the oil shipped to the US is found in conflict zones, and the increased pressure of developing countries demanding oil, the desire to permit alternative energy sites has decreased. Green groups have found that they can achieve large successes in court through lawsuits involving the purported violation of environmental laws. In the state of California an entire environmental institution has developed changing how project are handled and considered by the public. Additionally the push for preservationism, or an extreme form of conservation, has grown.

OUTCOMES

Currently, the future looks good for BrightSource Energy. They have the support of state and federal regulators. Their opposition, while vocal, has yet to file any suits or threaten any serious conflict with the company. There is no guarantee, however, that this will continue. The opposition could easily decide to file suit, or join an already organized national group. Issues regarding how the land is divided, how the desert tortoise is handled, or even threats to the viewshed, could be taken to court.

Conservations groups often benefit from filing and winning lawsuits. Not only do they gain the positive feelings that come with achieving a goal, or protecting nature and natural resources, but also they win attention from the case. Once these groups prove that they can handle themselves in the courtroom, they are more likely to attract donations of both time and money. Increased media attention from a case can also help these types of groups gain more followers.

BrightSource Energy could also gain much if the Ivanpah facility is a success. First, they could prove themselves to the local community with jobs and by providing extra protection to the area around their plant. Additionally, if their improved solar cell system is as effective they say it is, they will have increased demand for solar plants. Much of BrightSource Energy's success depends on available tax credits and subsidies. In the future, this could prove to be an Achilles' heel for the company again. There is no guarantee of continued federal funds and healthy tax credits.

FUTURE

Despite all of the rhetoric from both politicians and regulators, solar power faces severe levels of opposition. Even as the BLM has attempted to identify areas in the West where solar power could be sited, citizens and conservation interest groups have become more vocal in opposition to solar energy development. Thus, as politicians and regulators attempt to work to achieve generally approved of goals, conservation and environmental groups oppose them, at times the same groups that advocate alternative energy development.

6   Geothermal Energy

INTRODUCTION

The Modoc Plateau contains the Medicine Lake Caldera and the historical lands of the Pit River Tribe. The Pit River Tribe has a long tradition that surrounds the Caldera, and in particular the waters found therein. The area has long been a place viewed by various Native American tribes as having particular healing properties and the waters of particular interest. In recent years however, it has become clear that the area surrounding Medicine Lake has more to offer than just the historical and cultural importance placed on it by the Pit River Tribe. A geological study performed by Stanford University faculty, and sponsored by Calpine Corporation (a geothermal energy development firm), found that "[Medicine Lake] is perhaps the most promising, currently undeveloped, electrical-grade geothermal resource in the contiguous United States" (Hulen et. al., 2000). It is a region that has sufficient geothermal heat and water resources to provide substantial amounts of alternative energy. In fact, Calpine has submitted plans to develop two plants in the Caldera. Each would provide 49.9 megawatts of energy. Together that equals nearly 100 megawatts, or enough energy to power over 10,000 homes. Additionally, Calpine has stated that over the next forty-five years, in which they plan to develop geothermal energy sites in the area, they expect to see as much as 1,000 megawatts of energy provided (Pit River Tribe, 2011). In this chapter the story will be told of how Calpine was thwarted in their development through federal regulation by the Pit River Tribe Coalition.

HISTORY

One of the first uses of geothermal energy occurred over 10,000 years ago when early North Americans used hot springs for cooking (Energy Story, 2011). For centuries, naturally occurring hot springs have also been hotspots for relaxation. It hasn't been until recently, though, that geothermal power has been harnessed to heat buildings and power electrical grids. The first use of geothermal energy to power an engine was by Prince Piero Ginori Conti in Italy (Sanchez et al., 2011, p. 1). In 1904, he used steam to heat a small generator; eventually, he was able to use this steam technology to heat an entire power plant (Sanchez et al., 2011, p. 1).

The United States is currently the leading producer of geothermal energy, with the majority of this power—about half of the world's total geothermal production—(How Geothermal Energy Works, 2009; Energy Story, 2011). The earth's magma, a combination of hot and liquid rock under the earth's crust, is the source of geothermal power. This power contains roughly 50,000 times more energy than all the oil and natural gas in the entire world (How Geothermal Energy Works, 2009; Energy Story, 2011). This vast energy supply can only be harvested at seismically active hotspots, which reach down about 33,000 feet under the surface. The geothermal activity closer to the earth's surface can also be used for energy purposes; however, their power is more mild and can usually only be used for heating.

Geothermal energy is generated within the earth's crust, almost exclusively in the form of heat. This energy comes from two main sources; radioactive decay, which is continuously occurring, and residual energy from the creation of the planet. Radioactive decay contributes roughly eighty percent of the energy produced within the earth, while residual energy contributes much of the remaining twenty percent. Both forms of energy are released as thermal energy, which heats water near the surface. Traditionally the areas with the highest underground temperatures are located in regions with active or geologically young volcanoes. These so-called hot spots occur often at the boundaries of tectonic plates or areas where the crust is thin enough to let the heat through (How Geothermal Energy Works, 2009). The heat produced at these sites has heated homes and steam baths for centuries. In the late nineteenth and early twentieth century, advances in technology began to allow for the use of such heat to generate electricity. Since this time, the use of geothermal energy has expanded around the world, with some nations collecting over twenty percent of their power from this source. This expansion is not the case in the United States, however, where this environmentally friendly source of power is most widely utilized, yet still fails to meet even one percent of the nation's energy production. Overall, the energy produced by geothermal sources totals a mere 0.3% of the nation's annually produced energy (Geothermal energy technology and current status, 2002).

Geothermal energy is energy created and held within the Earth itself. This heat can be stored at any level of the earth, from mere miles below the surface to depths of over four thousand miles. Within the core, temperatures of nine thousand degrees Fahrenheit result in the conduction of heat into the surrounding rock found in the mantle and eventually the crust. The earth's thermal energy flows from the core to the surface at a rate of approximately 44.2 terawatts, and is replenished through the process of radioactive decay at a rate of roughly 30 terawatts (Rybach, 2007). Although these power rates are roughly double current energy consumption, the large majority of this power is not recoverable. The first assessment made by the U.S. Geological Survey estimated that geothermal sources across thirteen western states have the potential capacity to produce 8,000–73,000 MW, with a mean estimate of 33,000 MW (How Geothermal Energy Works, 2009) This would indicate that expansion into greater usage of geothermal energy has the potential to meet future energy demands.

Estimates indicate that geothermal energy is sustainable because the amount of recoverable energy is rather small in comparison to the total amount of energy created by the earth. This statement must be qualified by stating that particular areas can experience local depletion in the output of heat. Local depletion has been seen in some of the older sites, such as the Geysers in California, which have experienced reduced output. This reduction in output is largely due to the amount of time the Geysers have been tapped for their resources. Despite having the potential for over utilization as seen in all energy sources, geothermal is nonetheless an alternative, sustainable energy source.

The history of geothermal usage began in Paleolithic times when people used hot springs for bathing. As civilizations advanced, so did the use of geothermal steam. The first known spa appeared in China during the third century B.C. The Romans also made wide use of the hot springs found throughout their empire. The first record of geothermal use in heating systems appeared in the district heating system found in Chaudes-Aigues, France during the 1300's. Despite this wide use of geothermal energy for heating purposes, the use of this heat to produce electricity did not come about until the mid-twentieth century. The first successful geothermal power plant in the United States opened in 1960 at The Geysers in California (Lund, 2004). The Geysers is also the largest geothermal system in operation to date. It now supports twenty-six power plants, with a capacity of over 2,000 MW (How Geothermal Energy Works, 2009).

Currently, there are twenty-four countries with functioning geothermal electrical power plants. Annually, the energy produced by these twenty-four countries could power approximately 12 million American households (How Geothermal Energy Works, 2009). Of these nations, the United States is the leader in geothermal electricity, with the installed capacity to generate approximately 3,086 MW of electricity (Holm et al., 2010). Despite its lead in total energy production, geothermal energy represents only a third of a percent (0.3%) of the annual national production in the U.S. This level of use starkly contrasts nations such as the Philippines (27%), El Salvador (25%), and Iceland (30%) (Holm et al., 2010). Traditionally, geothermal electrical power producers have built plants on the edges of tectonic plates, but advances in drilling and extraction have opened a much wider geographical region for geothermal extraction.

Geothermal energy is captured in many different ways, but the most common method involves tapping into naturally occurring hydrothermal convection systems. These systems are areas where cooler water seeps into the earth's crust, heats up, and then rises to the surface. When the heated water rises to the surface, it is relatively simple to capture the steam and use it to drive electric generators (How Geothermal Energy Works, 2009). Geothermal power plants drill their own holes into the rock to more effectively capture the released steam. There are currently three designs that pull hot water and steam from the ground, utilize it, and then return it to the ground. In the simplest design, steam goes directly through a turbine to a condenser and is then returned to the earth as water. Another method depressurizes very hot water and "flashes" it into steam that drives a turbine. In a third method, known as the binary system, hot water passes through a heat exchanger, where it heats a second liquid in a closed loop. The liquid in the closed loop boils and is converted into steam to run the turbine. Binary systems typically use liquids like isobutane as a secondary liquid. The method employed depends largely upon resources available at each individual site (How Geothermal Energy Works, 2009).

Although generally considered a clean, alternative energy source, geothermal energy still presents some environmental concerns. For example, the environmental impacts of geothermal electrical generation are greater than other areas of alternative energy because gases released during production contain carbon dioxide, hydrogen sulfide, methane, and ammonia.

The environmental impact of geothermal energy is not limited to released gases. The water from geothermal sources often contains traces of toxic chemicals as mercury, arsenic, and boron. These chemicals are obviously harmful if released, but modern techniques of injecting the cooled fluids back into the earth both stimulates production and reduces the environmental impacts associated with these chemicals. Other environmental impacts include land instability, as seen during a developmental drilling operation near Basel, Switzerland. This project was suspended due to over 10,000 seismic events that measured up to 3.4 upon the Richter scale during the first six days of operation (Deichmann, 2007).

Relative to conventional energy production, geothermal plants offer an environmentally friendly alternative. One of the benefits of geothermal electrical generation is minimal land and freshwater use. Geothermal plants use roughly 1.4 square mile per gigawatt (GW) of electrical production, compared to twelve square miles for coal burning power plants. Geothermal power plants also use 5.3 gallons of freshwater per megawatt hour as opposed to over 260 gallons required for nuclear, coal, or oil (Lund, 2007).

The Geothermal Technologies Program (GTP), under the direction of the U.S. Department of Energy, is developing technologies that will assist with location, access, and development of geothermal energy. The goals of the program include lowering the risks and costs of development and exploration for new sources of geothermal sites, lowering the cost of electricity, and developing at least 30 GW of energy from currently undiscovered hydrothermal resources (Geothermal Technologies Program, n.d.). Ultimately, the GTP has set the goal to lower the price of hydrothermal electricity to six cents per kilowatt-hour (Geothermal Technologies Program, n.d.). To accomplish this goal, the GTP has established several strategies, such as: increasing the speed of innovation as it relates to the development of geothermal energy, reducing the cost and risks of exploration and development, and beginning a system of validation in order to increase investor confidence (Geothermal Tomorrow, 2008). The GTP is currently working with industry, academia, and the U.S. Department of Energy's national laboratories to continue advances in technologies and further expansion of its usage as a major energy source. The program met considerable support from politicians in Washington who, as part of the 2009 budget, awarded approximately $80 million to continue the program's efforts after a struggle for funding during the previous two years (Geothermal Technologies Program, n.d.). As demonstrated by these initiatives, the United States is attempting to expand geothermal power generation to play a more significant role in the country's energy production.

The majority of U.S. geothermal sites can be found in the thirteen western states, the most prominent including California, Nevada, Oregon, Washington, Alaska, Utah, Wyoming, and Montana. The Pacific Coast states are on the list because of their location on the "ring of fire". The "ring of fire" is an area where the Pacific Ocean tectonic plates meet the North American plate and other continental tectonic plates.

There are many reasons why little geothermal energy is developed. Primarily, energy production from fossil fuels is cheaper, less risky, and more entrenched. Despite its numerous advantages, geothermal energy also has a number of drawbacks. Because geothermal energy has a limited impact on the energy market in the U.S., there is a shortage of the infrastructure, technologies, and skilled workers in the field. High installation costs and sharp financial risks for investors are additional obstacles preventing further development of geothermal energy. Companies that are able to fund high installation prices face additional challenges in hiring a certified installer and finding skilled staff to manage the site. The chances of steam depletion, as discussed earlier, can be quite devastating to smaller operations or recently created companies. These aspects of geothermal energy lead to increased market risks, and fewer investments as a result (Advantages and disadvantages of geothermal energy, n.d.).

Geothermal energy has applications beyond electricity production. Since the thirteenth century in France, people have harnessed geothermal energy to heat houses and other structures. The use of geothermal energy for heat came to the United States in the late nineteenth century to Boise, Idaho. This effective method for heating buildings has proven quite effective in cold weather areas such as Alaska, where geothermal has replaced more traditional heat sources such as fossil fuels. Other non-electrical uses of geothermal energy include use to heat greenhouses and various industrial processes. As of 2000, the total usage of geothermal energy in extra-electric applications totaled about 15.14 MW (Geothermal energy technology and current status, 2002).

Many methods exist for harnessing the earth's heat beyond the traditional method of dry/wet rock extraction. The notion of using magma bodies and geopressured reservoirs is relatively new, but additional research into these new methods is required before industrial application. Potential for breakthroughs and further advances in the field of geothermal research and application provide hope for the future.

Regulations of geothermal energy have been somewhat recent in their enactment due largely to the recent movement towards exploration and expansion of geothermal resources. The first law that bears considerable mention is the Geothermal Steam Act, which was essentially the first law directed at the usage of land for geothermal energy. Congress passed this law in 1970, with the objective of allowing leases of land containing geothermal resources. This law excluded any lands located within the National Park system as well as any land, which was prohibited from leasing by the Mineral Leasing Act of 1920 (Kubiszewski, 2008).

Several laws that deal directly with Indian lands also have direct impact upon the geothermal industry. The two laws that deal directly with leasing upon tribal lands are the Indian Mineral Leasing Act of 1938 and the Indian Mineral Development Act of 1982. The first of these laws was passed to allow for the leasing of tribal lands, with the consent of the specific tribe in question. The second law allows a tribe to enter into an energy development agreement with the approval of the Department of the Interior. These two laws are subject to several regulations imposed by the Bureau of Indian Affairs, notably 25 CFR 211 and 25 CFR 225 (Laws and Regulations Applicable to Geothermal Energy Development, n.d.). CFR 211 specifically deals with guaranteeing that any mineral development upon Indian lands will be done in a manner that benefits the tribe to the maximum level possible, and minimizes adverse environmental or cultural impacts (Amar, n.d.). Additionally there has been further restriction placed upon land available for lease and lessee qualifications under 43 CFR 3200, which edited several regulations set forth under the Geothermal Steam Act (Part 3200—Geothermal Resource Leasing, n.d).

THE PROCESS

There are four main ways that geothermal energy is used to create electricity: Dry Steam, Flash Steam, Binary Cycle, and the Combined Cycle.

The dry steam method is used with the hottest category of water and was the first method utilized to create electricity from geothermal sources. To power the turbines, steam is funneled directly from the vent to the generation machinery. There it is used to turn the turbines and power the machinery.

The flash steam method is the most commonly used today, and uses water that is nearly as hot as the dry steam method, being at temperatures greater than 360 °F, but maintains a liquid form due to the pressure surrounding it. To maintain that liquidity, the water is pumped under high pressure to the generation equipment. Once there, the pressure is suddenly reduced and the water flashes into steam. The steam created here is used to drive the turbines just as the steam in the previous method was. The water that wasn't hot enough to be flashed into steam is sent back down into the earth.

The third method of harnessing electricity from geothermal energy is done in a bivalve cycle. This water is not hot enough to flash into steam. To bypass that problem, the water is used to heat another "working liquid," which has a lower boiling point than water. Thus, this liquid will turn into steam at a lower temperature. This water substitute then turns to a gas and moves into the generators. This liquid is then reused once cooled. This bivalve system is a closed system that doesn't put pollutants into the atmosphere. Naturally, this binary system is quite suitable when the geothermal hot spot isn't warm enough to power turbines with water alone (How Geothermal Energy Works, 2009). Since this is the case with most hot spots, this system has the most potential. The binary system is also beneficial to the environment because there are no emissions, as there are with the other options.

In the final system, the combined cycle, the flash and binary systems are combined into one system (Blodgett & Slack, 2009). Here the less pressurized steam is funneled through a 'flash' conversion and transformed into electricity. The remaining steam is pushed through a backpressure steam turbine and then through a binary system (Blodgett & Slack, 2009).

Geothermal is seen as an important tool for future energy production. Benefits include an almost constant base-load, or a constant influx of steam to turn into energy. Geothermal energy is the only alternative energy that provide this benefit—making it comparable to coal.

This constant base-load is particularly the case with Enhanced Geothermal Systems (EGS). EGSs provide a new opportunity to access the heat of the earth commercially. To this point, geothermal energy has largely been utilized in hot spots and geyser areas. These are places where hot water comes to the surface naturally or is easily accessible. EGS, or dry rock as it is often called, is used to access heat from a wide variety of places. A hole is drilled three to five kilometers deep into the crust of the earth and cold water is sent down to fracture already existent cracks in the rock (Tester et al., 2006). This hot rock then vaporizes the water.

Additional holes are dug to siphon this heated and pressurized water back to the surface for utilization at a geothermal facility. As previously stated, this siphoning drastically increases the area where geothermal energy can be found. However, one major drawback of this approach is the increased chance of an earthquake (How Geothermal Energy Works, 2009). Co-generation plants, another promising future of geothermal energy, combine the power of natural gas or oil along with the steam or hot water that is generally found in these areas. Often the water found in these areas is under a significant amount of pressure, perfect for use in generating electricity from both geothermal heat and the petroleum derivatives there (How Geothermal Energy Works, 2009).

Another option, currently running in Iceland and Germany, is the Mixed Working Fluid System (Blodgett & Slack, 2009, p. 12). It utilizes a mix of both ammonia and water in a flash system. The mixture of these two liquids achieves a higher efficiency than is possible with either independently (Blodgett & Slack, 2009, p. 12). This system is also able to run at a lower temperature than other systems, allowing for more widespread implementation. Construction sites distant from urban centers have also begun to rely on geothermal energy as a means for on-site electricity (Blodgett & Slack, 2009, p. 13).

Chemicals such as CO_2 and hydrogen sulfide, arsenic and other minerals are released with the steam (How Geothermal Energy Works, 2009). If you have ever visited Yellowstone National Park and smelled the geysers there, you know they come with a rotten eggs smell, a byproduct of hydrogen sulfide. The emissions produced from the most "damaging" of the plants, dry steam plants, is still significantly less damaging than any fossil fuel plant. Geothermal plants have even been known to improve air quality, such as The Geysers plant in California (Blodgett & Slack, 2009, p. 31). There have yet to be any instances of geothermal extraction activity contaminating groundwater (Blodgett & Slack, 2009, p. 48). Additionally, many people may worry about the smoke they see geothermal plants emitting, however this is only steam (Blodgett & Slack, 2009, p. 47); this energy source is also less land intensive than any other energy source, both dirty and clean (Blodgett & Slack, 2009, p. 33).

When it comes to siting geothermal plants, there are many conditions that should be met for an optimum location; mainly, the geologic formation being used should be very stable. There should be little chance or history of landslides, seismic events, or caving (Environmental Impacts, Attributes, and Feasibility Criteria, n.d., p. 17). There must be a way to reuse waste-water and heat (Environmental Impacts, Attributes, and Feasibility Criteria, n.d., 17). Other feasibility conditions must be met as well. Naturally, it must be located near either a demand for electricity or heat (Environmental Impacts, Attributes, and Feasibility Criteria, n.d., 17). If the site is too distant from either of these, then it must be close to a transmission infrastructure (Environmental Impacts, Attributes, and Feasibility Criteria, n.d., 17).

REGULATIONS

Plans for many geothermal plants have been thwarted and eventually stopped because of litigation empowered by federal regulation. Many acts of Congress came into play in these cases, including the National Environmental Policy Act (NEPA), the National Historic Preservation Act (NHPA) and the Native American Religious Freedom Act of 1979.

The Geothermal Steam Act was passed in 1970, with the goals of developing geothermal energy in the U.S. and decreasing dependence on fossil fuel energy sources that were becoming increasingly volatile. Pursuant to this act, the Department of the Interior issued a nationwide programmatic Environmental Impact Statement (EIS) in 1973. The department wanted a general idea of what geothermal energy availability and restriction based on its location. This EIS was intended to be a generic analysis and "provided for tiered environmental review because of the wide geographical distribution of potentially affected lands." The EIS also indicated "specific details will be identified, evaluated and described in the environmental analysis record prepared for each lease area prior to any leasing action". An additional stipulation of the EIS was that more environmental evaluation would be necessary, should construction or planning for new power plants or related facilities occur (*Pit River Tribe v. U.S. Forest Service*, 2006).

CASE STUDY—TELEPHONE FLATS AND FOURMILE HILL

Among the areas investigated in the EIS was the Medicine Lake Caldera in northern California. Finding this region to have exceptional geothermal energy capabilities, the Secretary of Interior designated the Medicine Lake Highlands as the *Glass Mountain Known Geothermal Resource Area* (KGRA). Subsequent to that declaration, two Environmental Assessments were conducted in 1981 and 1984. The first of these assessments did not address cultural or tribal impacts of leasing and development, though it pointed out that "a decision to lease carries with it the right to develop a discovered resource, subject to the limitations of the lease," meaning that the lands would be developed; this in turn proved to drive the direction taken with these lands in subsequent decades. The limitations outlined were crucial in development considerations as they were the environmental safeguards required throughout the process, including more EAs and EISs (*Pit River v. U.S. Forest Service*, Lewis, 2006).

The second EA in 1984 was also significant because it was the first to consider the "potential effects of leasing activity on cultural, recreational and spiritual aspects of certain features of Medicine Lake" (*Pit River v. U.S. Forest Service*, Lewis, 2006). During this investigation, it was reaffirmed and published "that the area remained culturally significant to the modern-day Native Americans and that the American Indian Religious Freedom Act of 1978 continued dialogue with both Native American tribes and individuals in order to protect sites important to cultural traditions" (*Pit River v. U.S. Forest Service*, Lewis, 2006). In addition, they found that any development or change of natural features would likely interfere with Native American culture in the Medicine Lake region. The assessment also acknowledged the potential historical significance of the area but concluded that even acceptance to the National Historic Registry would not be enough of a reason to stop scientific exploration and excavation. To address these potential impacts, the Bureau included a table of mitigating measures. As a boundary of environmental protection, the EA still required that the lessee prepare "an operating plan for subsequent activities in exploration, development and operation of the lease" that would include additional environmental analysis and approval." (*Pit River v. U.S. Forest Service*, Lewis, 2006). In short, the EA discovered that the land was much more significant to the Native American people than was originally thought.

THE PLAYERS

The main participants in this controversy are the Pit River Tribe and the Calpine Corporation. Siding with the Pit River Tribe are the Native Coalition for Medicine Lake Highlands Defense and the Mount Shasta Bioregional Ecology Center. Calpine's allies include the United States Forest Service, the Advisory Council on Historic Preservation, and the United States Bureau of Land Management.

PIT RIVER TRIBE

The Pit River Tribe is a sovereign Indian Tribe, federally recognized and consisting of eleven autonomous bands or groups. As we noted previously, Medicine Lake holds special significance to the Pit River Tribe. They believe that the creator and his son bathed themselves there after creating the earth and in the process, endowed it with healing powers. Though the lake is no longer within the tribal boundaries, their historical territory includes Medicine Lake and the surrounding Highlands. The Tribe's current claim to the area is the history they have shared with it and its significance to their culture that reportedly has dating back 10,000 years or more. This history was established by religious and cultural use; the land served as a training ground for ceremonies such as "vision quests, religious prayers and teaching, traditional shaman and doctoring practices, life cycle ceremonies, collection of traditional foods, medicines and materials, spiritual renewal and quiet contemplation" (Ninth Circuit Reverses Lower Court Ruling, 2006). The Tribe continues to use the region for these activities and more.

The Pit River Tribe is not alone in its sacred view of Medicine Lake. There are many tribes that view this land as sacred and special. We noted earlier that nearly every tribe between the Pacific Ocean and the Rocky Mountains sent their medicine men there to train (Pit River Tribe, 2011). Most of the other tribes however, have remained passive on the issue. For example, the Shasta Tribe was deeply engaged in pursuing government recognition. Calpine Corporation promised to help with their lobbying efforts and promised college scholarships and jobs at Calpine's new facility to young tribe members. As such, the Shasta Tribe found greater benefit in staying out of the issue. With these offers of assistance also flowing to them, the Klamath and Modoc Tribes (the region's other main tribes) also opted to not interfere in the geothermal issue, leaving the Pit River Tribe as the primary native nation actively opposing the project (Bailey, 2002).

CALPINE CORPORATION

Acquiring its name from the combination of its California location and alpine funding (from Electrowatt Ltd., a Swiss company), Calpine was founded in 1984 by Peter Cartwright and two other former executives of the San Jose office of Gibbs and Gall. The original purpose and mission of the company was to provide a service of "engineering, management, finance and maintenance services" to the independent power production industry. Calpine turned its first profit after two years but saw greater opportunities for growth in pursuing a focus more akin to what their clients were doing-plant development (Calpine Corporation, n.d.).

With that realization, Calpine set out to compete in its new role as a plant developer and operator. Now the focus was on "acquisition, development, ownership, operation, and maintenance of gas-fired and geothermal power generation facilities" (Calpine Corporation, n.d.). Starting small, the company quickly posted profits of $40 million a year and provided approximately 297 megawatts of electricity. This appears to be a large sum, until it is compared with the power generation industry as a whole, which brings in $200 billion a year. Though Calpine had a small market share overall, it was recognized very early as having some of the best geothermal energy facilities in the industry, ranking among the top four or five in the early 1990's (Calpine Corporation, 2010).

The 1990's marked a period of sharp deregulation in the energy business. As an independent power company, Calpine was in an excellent position to capitalize on this change. The care that had been taken to vertically integrate the company proved to also aid in using the deregulation to their advantage. They were able to keep costs low and offer very competitive rates. Calpine had foreseen and prepared for this deregulation. As such, they were able to capitalize on this boon in a way their competitors could not. They also took gambles with plant equipment purchases, often buying more than they had facilities for. This proved to work out in a lucrative way, as they were able to stay ahead of the steep demand and buy low to sell high.

THE LEASES

As previously discussed, the National Environmental Policy Act requires environmental assessments and statements to be performed prior to any leasing. These assessments had been done with the limitation that should any development be expected, further assessments in greater detail would be needed. With that understanding, sites began to be leased in the Medicine Lake Caldera in 1988. One of them, the Fourmile Hill site, went to Calpine Corporation. These leases were for minimal activity and had a time limit of ten years. As we described earlier, no additional environmental or cultural analysis was taken with the issuances of these leases. This was not seen as necessary, as these were simply exploratory leases. Should further development be planned and requested, another set of analyses would be required.

It wasn't until September of 1996 that a plan was submitted by Calpine to explore a geothermal site in the Fourmile Hill area. With that plan proposed, the agencies involved namely the US Forest Service, went about preparing an EIS. This EIS was met with a great deal of vocal disdain from the public and Pit River Tribe. After four years of deliberation and investigation however, the Forest Service approved Calpine's development plan.

In May of 1998, the ten-year lease was up. Not much had been done with the site but the Bureau did in fact extend Calpine's leases. This extension was done for five additional years. No further environmental assessments were done prior to the extension of this lease. In fact it was September, just a few months after the lease extension, "the agencies issued a final EIS for the Fourmile Hill Plant (1998 EIS) that included different configurations for the facilities" (*Pit River Tribe v. U.S. Forest Service,* Lewis, 2006). They chose an alternative that seemed not to have much of an impact on the surrounding areas or cultures. A "no action" option was barely considered in the statement. But it truthfully wasn't really a viable option as development was the only course to take, with the leases now out of federal hands.

A year later, the Keeper of National Register of Historic Places issued a report determining that the Medicine Lake Caldera was eligible for listing on the National Register, and the lake was then added to the record. While the Fourmile Hill facility was not currently within the Medicine Lake historic region a call was made for investigations on sites near the new historic area to be evaluated for more possible additions to the register. To aid in these investigations, a moratorium was placed on the Fourmile Hill site, and no development could take place during those five years while the environmental evaluation took place.

In 2001, Calpine purchased the other Medicine Lake lease, Telephone Flats. This newly acquired lease was located squarely within the Medicine Lake Historic region. That being the case, any development on the land was also stalled, pending environmental and historical analysis. Hence, developing the new site was postponed indefinitely. Calpine proceeded to sue the government for $100 million for the interference. Following a serious lobbying effort in Washington, the moratorium was lifted on Fourmile Hill and Telephone Flats was also taken off suspension, with the requirement of a shift in power lines so as not to interfere with sacred lands. With this result, Calpine dropped its suit against the government. Calpine noted, however, that they would be willing to reopen the case, should their ownership of the lease be inconvenienced again.

In addition to removal from moratorium and suspension, the leases were granted a forty-year extension. This was done without any further EISs performed. Another issue that the agencies failed to follow was the fact that Telephone Flats was now a part of the Medicine Lake Historic site and Fourmile Hill was under consideration. As such, a Historic Analysis needed to take place prior to the extension of these leases. This might have passed by unnoticed, but the Native Americans (particularly the Pit River Tribe), were not properly consulted in the issuance and extensions of these leases, nor were they consulted on the entailed development.

THE LITIGATION PROCESS

To advocate their case, the Pit River Tribe and other interested groups recruited the Stanford Environmental Law Clinic. The clinic consists of ten to fifteen law students and their mentors who advocate on behalf of environmental, non-profit groups. They have successfully litigated many cases ranging from protecting the Joshua Tree National Park and the Mojave Desert tortoise to National Forest Fire Management policy and post-Katrina wetlands destruction.

The Calpine plants had been approved at the federal level so the Pit River coalition took the federal agencies involved, as well as Calpine, to court.

They contested that both the five-year and forty-year lease extensions were in violation of NEPA. Litigation continued for both sites until December 20, 2005, when Calpine Corporation declared Chapter 11 Bankruptcy. With over $26 billion in assets and an inability to operate, Calpine became one of the largest companies in history to declare bankruptcy, the ninth largest in the U.S., ranking nineteenth in the world (Ross, 2010). Though they re-emerged in January 2008, their bankruptcy effectively stalled the Telephone Flats project. But they were able to continue attempting to develop the original Fourmile Hill site.

The litigants in the case included The Pit River Tribe, the Native Coalition for Medicine Lake Highlands Defense and the Mount Shasta Bioregional Ecology Center. Yet, on the court record, they refer to all groups as Pit River. We will follow the same pattern to simplify the list of characters. On the defending side, there was the United States Forest Service, the Advisory Council on Historic Preservation, the U.S. Bureau of Land Management and Calpine Corporation. In recounting the story of the court decisions, these participants will be collectively referred to as, "the agencies."

The Pit River Tribe coalition viewed federal regulation as an ally in stopping this geothermal development by Calpine. The acts that were put forward as being trespassed against were The National Environmental Policy Act, the National Forest Management Act, the Endangered Species Act, the National Historic Preservation Act (NHPA), the Federal Land Policy and Management Act and the Freedom of Information Act. The Eastern District of California ruled in favor of Calpine and the federal agencies. The case then went to the 9th Circuit Court of Appeals, where the Pit River Tribe got its much-needed break. The court evaluated the case under the lens of the Administrative Procedure Act (Pit River v. U.S. Forest Service, Lewis, 2006), and in the Court proceedings, the case for the federal agencies and Calpine quite literally dissolved.

THE 9TH CIRCUIT

When the litigation reached the 9th circuit in 2006 one of the core questions facing the court was whether or not the case was valid and if the issues raised could be considered through litigation. At the core of these questions was whether the Pit River Tribe and their allies had legal standing to engage in litigation. The court first had to resolve the question of standing before it could evaluate whether an EIS was required and if the completed EISs were inadequate.

At the filing of the lawsuit, the statute of limitations had expired on the 1988 lease agreements. As such, Pit River focused their energy on the 1998 lease extension. The core argument of the case was that a proper environmental review had not taken place prior to the 1998 extension, which, they claimed violated NEPA and NHPA. The agencies held that they had they had performed four assessments: the 1973 EIS, both the 1981 and 1984 EAs and another EIS that was completed just a few months after the 1998 lease extension. The agencies contended that they had done more than enough in their environmental assessments for development to move forward.

QUESTION ONE

As previously noted, the court needed to first establish whether or not the case was valid. To do this, two questions needed to be answered—first, the court needed to establish whether or not the plaintiff (the Pit River Tribe) had Article III standing, or in other words- something the court has juris- diction to hear and resolve. "Standing to sue" rests on three requirements: an understanding of whether or not the plaintiff suffered an "injury in fact"; second, that the injury can be traced to the defendant, and third, the issue must be something that can be properly redressed through a favorable deci- sion by the court. The second overarching question posed to the court was whether the 2005 amendments to the Geothermal Steam Act affected the claims placed on these sites (*Pit River Tribe v. U.S. Forest Service*, 2006).

On the subject of "injury in fact", it was argued by the agencies that they were guilty of no such injury toward Pit River. They asserted that they had acted in accordance with the law and had not overstepped their boundaries. Using precedence the court held that that the plaintiff who claims procedural injury (in this case the Pit River Tribe) must show that the procedures that were trespassed against them were designed to protect the plaintiff (*Pit River Tribe v. U.S. Forest Service*, 2006). Essentially the court required that the Pit River Tribe prove their relationship with the area before they could claim a concrete interest and show a legitimate injury, should harm come to it.

Ironically enough, the proof needed to establish the standing of the Pit River Tribe's connection to the area under question came from the agen- cies' very own 1996 ethnographic report. It reads, "Previous research has indicated that the Medicine Lake Highland and Timber Mountain areas have long been recognized by non-tribal scholars as traditional cultural properties of . . . the Pit River Nation." The report continues by explaining that the Tribe used these lands for religious and cultural ceremonies. They have had this relationship with the region for "countless generations." With this written statement, "injury in fact" was assured for the Pit River Tribe (*Pit River Tribe v. U.S. Forest Service*, 2006). Thus, the first requirement had been met.

The agencies further argued Pit River's Article III standing by stating that the leases were already given in 1998 and again in 2002. To the agen- cies, this meant that that there was no redress ability available to the Pit River Tribe toward the 1998 lease because the 2002 lease had surpassed it. In addition, they held that even if redress ability was needed, it would be provided by the benefits of the plant in the area. The court rejected this logic on the grounds that if the court holds the Environmental review to be less than adequate, they then had the power to either invalidate the 1998 lease (which would also remove the 2002 lease) or they could take control over any surface activity until the agencies fully comply with NEPA (*Pit River Tribe v. U.S. Forest Service*, 2006). Hence, under this analysis by the 9th Circuit Court, the Pit River Tribe indeed was found to have Article III standing. They had incurred injury, stemming from the agencies' activities and the issue could be remedied by a favorable court ruling.

QUESTION TWO

The court now needed to answer one other question to determine whether or not they could hear this case. This question required an understanding of what effects the 2005 amendments to the Geothermal Steam Act had on the claims placed on these sites. If it did have an impact, then that might explain how to handle the situation. To understand the nature of the amendment effects on the case, "[The Court] adopted the test from Landgraf v. USI Film Products," which is, "if Congress expressly states that an amendment applies retroactively then the matter is settled" (*Pit River Tribe v. U.S. Forest Service*, Lewis, 2006). Yet, if that expressed statement is not present, then it is to be decided by the court and in all but extreme cases (depending on whether it would apply new duties on a party) it is not applied retroactively.

After carefully evaluating the amendments to the act, the court determined there was no such express statement. As such, the decision was theirs. They decided that should the act be applied retroactively, it would impose new duties on Calpine including new minimum work and payment requirements. The court ruled that the act did not apply retroactively (*Pit River v. U.S. Forest Service*, 2006). With this conclusion, the second question was answered and now the case could be heard and ruled upon.

THE CASE

The necessary questions are answered. There is a case. Because the statute of limitations had run its course on the 1988 lease, Pit River focused their energy on the 1998 lease extension. The Tribe argued that a proper environmental review had not taken place prior to the 1998 extension, violating both NEPA and NHPA regulations. In its defense, the agency argued that they had performed four assessments: the 1973 EIS, the 1981 and 1984 EAs, and another EIS just a few months after the 1998 lease extension. The agencies maintained that the final EIS was the validation for the lease agreements and the existence of the power plant.

An important consideration to note is that NEPA "requir[es] agencies to (1) consider the environmental impacts of their proposed actions and (2) inform the public that they (the agencies) considered environmental concerns in their decision-making process" (Luther, 2005). With that expectation, the 9th Circuit Court needed to be fully assured that the agency took a "hard look" at any and all of the environmental consequences of their decision. This level of detailed assurance was needed so as to reaffirm that this amount of precaution and regulation in place is done to dictate procedural safeguards rather than a certain result (*Pit River Tribe v. U.S. Forest Service*, 2006).

DID ANOTHER EIS NEED TO TAKE PLACE?

Pit River argued that even though the 1998 EIS was published just months after the lease renewal, it was not valid because it did not follow the prescribed path outlined in NEPA—that the EIS be completed prior to the lease extension. The court again had precedence to reference with respect to this particular argument. Previous cases had not required that an EIS be conducted if the government had not irretrievably committed resources. For example, if the government maintains control over how much lumber can be harvested on land they have leased, or if they maintain control over what drilling can take place for oil, or what surface area can be interacted with, then they still have ownership. With that ownership, the land is not irretrievably given to the lessee. Therefore, an investigation needed to be made to find what level of ownership the agencies ceded to Calpine in the lease agreements (*Pit River Tribe v. U.S. Forest Service*, 2006).

The result of that investigation yielded some useful information; in the words of both the 1988 and 1998 leases, Calpine was granted "the exclusive right to drill for, extract, produce, remove, utilize, sell and dispose of the geothermal resources in the [leased] lands." Looking even further, the court found no reservations included in the lease for the agencies to have an absolute right to deny exploitation of resources. The only relevant factor they seem to have is the ability to limit development "when not consistent with the lease rights granted" (*Pit River Tribe v. U.S. Forest Service*, 2006).

In the agency's own reports, the court found the evidence that it needed in a briefing paper and the Record of Decision (ROD) for the Fourmile Plant. The briefing paper revealed that solicitors from the Department of the Interior had advised the agencies, "denial of the Projects would be a taking of private property rights associated with the leases . . . The decision makers *would like to have the authority* to deny the geothermal projects, which may require compensation to the leaseholders for the taking." It was found to be heavily implied by this portion of the brief, particularly the phrase, "would like to have the authority," that the agencies did not have the authority to deny these projects. In addition, the ROD recognized that Calpine saw the leases as granting an absolute right to develop. It is even admitted in the ROD that Executive Order 13007 (entitled Indian Sacred Sites) was rendered without effect, as Calpine *could not* be denied its vested rights as a leaseholder (*Pit River Tribe v. U.S. Forest Service*, 2006).

In other words, although the agencies believed they had the ability and power to deny any projects, the effect of the lease was that they in fact did not. If they had tried to deny any project, they would also be denying Calpine's vested rights as a leaseholder. Essentially, they would be interfering with private property without the proper authority to do so. With this level of control given up, an EIS is required prior to any lease agreements signed.

THE 1973 EIS

As we noted earlier, the agencies put forward the claim that the 1973 programmatic EIS and the 1981 and 1984 EAs were sufficient grounds to move forward on the leases and extensions. The court however rejected the 1973 argument on the grounds that it was a general EIS, only sufficient for a casual or exploratory lease. It did not apply to instances when leases would be irreversibly and irretrievably committed to commercial development. Though a previous court hearing in 1978 had given permission for these "casual" leases to be issued, eve that court acknowledged that should further development be intended, a proper supplemental EIS should be performed under the complete requirements of NEPA law (*Pit River Tribe v. U.S. Forest Service*, Lewis, 2006). On these grounds, the court also denied the 1981 and 1984 EAs, deeming them insufficient because they only dealt with the issue of leases and casual use exploration—not geothermal development.

The agencies then contended that the 1998 lease extension simply maintained the status quo and did not have any effect on the nature of the area and did not require a separate assessment to be issued. If the status quo were truly maintained and nature were left unaffected, that argument would be true and the lease extension would not require an EIS (*Kootenai Tribe of Idaho v. Veneman* in *Pit River Tribe v. U.S. Forest Service*, 2006). The court held, however, that the lease extensions were not merely a continuation of the status quo.

The final defense raised by the agency was that the 1998 EIS for Fourmile Hill plant was sufficient grounds for the proper redress to the Pit River Tribe. The court held that "The purpose of an EIS is to apprise decision makers of the disruptive environmental effects that may flow from their decisions at a time when they retain a maximum range of options" (*Pit River Tribe v. U.S. Forest Service*, 2006). The actual law states that "agencies shall integrate the NEPA process with other planning at the earliest possible time to ensure that planning and decisions reflect environmental values, to avoid delays later in the process and to head off potential conflicts" (Protection of Environment, 2012). The "no action" alternative is an important thing to consider when conducting an EIS. This particular EIS had been issued a few months after the lease had already been extended. This made it both untimely and unable to address the "issue of whether the land should be leased at all." Instead it discussed the configurations of the proposed power plant. This exhibited bias in the EIS, which the court viewed as invalidating the EIS (*Pit River Tribe v. U.S. Forest Service*, Lewis, 2006).

THE COURT'S CONCLUSION

Earlier, we discussed that the agencies needed to prove that they had taken a "hard look" in extending the leases and had met the procedural requirements under NEPA. The court concluded that such a thorough examination had not taken place. The court determined that "the 1998 lease extensions—and the entire Fourmile Hill Plant approval process for development of the invalid lease rights—violated NEPA" (*Pit River Tribe v. U.S. Forest Service*, 2006).

Despite this holding, NEPA wasn't the only act that was examined during the proceedings. The National Historic Preservation Act also was reviewed. The "NHPA is similar to NEPA, except it requires consideration of historic sites rather than the environment" *United States v. 0.95 Acres of Land*, 994 F.2d 696, 698 (9th Cir.1993) (*Pit River Tribe v. U.S. Forest Service*, 2006). "The Pit River Tribe also claimed that the agencies violated the NHPA by not identifying traditional cultural properties prior to the extension of the leases" (*Pit River Tribe v. U.S. Forest Service*, Lewis, 2006). The District Court in California had upheld the agencies' claim, saying that the 1998 EIS was sufficient to provide the relief sought by the Tribe and that the lease extensions were not under NHPA because it was simply maintaining the status quo. The 9th Circuit, however, concluded that "The NHPA involves a series of measures designed to encourage preservation of sites and structures of historic, architectural or cultural significance" (*San Carlos Apache Tribe v. United States* in *Pit River Tribe v. U.S. Forest Service*, 2006). They continued, "when an undertaking may affect properties of historic value to an Indian tribe on non-Indian lands, the consulting parties shall afford such tribe the opportunity to participate as interested persons" (Muckleshoot Indian Tribe v. U.S. Forest Serv., 177 F.3d 800, 806 (9th Cir.1999), quoting 36 C.F.R. § 800.1(c)(2)(iii). It had been established, through both the EIS and EAs, that this land was of cultural significance to the Indian Nation. The court concluded that no consultation had taken place. The agencies defense was that no NHPA analysis was required and if it had been that later analysis solved the problem. The court disagreed and said these analyses were required and could not be redressed by later analysis because the question at hand was whether or not the land should have been leased at all. The court held that NHPA, too, had been violated. The court held that:

"The agencies violated their duties under NEPA and NHPA and their fiduciary duty to the Pit River Tribe by failing to complete an Environmental Impact Statement before extending Calpine's leases in 1998. Hence, both the five-year lease extensions and the subsequent forty-year extensions must be undone. The rest of the project approval process, including the 1998 EIS, was premised on Calpine's possession of a valid right to develop the land and therefore, must be set aside" (*Pit River Tribe v. U.S. Forest Service*, 2006).

One of the most important realities that this case illustrates is the fact that there are multiple chokepoints that can be used to stop development. In this case, environmental, historical, and issues surrounding Native American rights had to be addressed in order for this plant to be constructed and to operate. Further because the agencies did not follow the exact procedures outlined by NEPA and NHPA, the result might have permanently cost Calpine these sites.

THE CURRENT STATE OF FOURMILE

In 2010, the characters in our story again found themselves in the 9th Circuit Court. The decision was now whether Calpine would have to apply for a new lease or if it could proceed as if applying for a lease extension. It was decided, to protect future lessees from having their leases removed for lawsuits, that it would be treated as a lease extension. Calpine is applying for the lease extension and the Pit River Coalition remains a roadblock to any future development from taking place.

ANALYSIS

In this policy arena the Pit River Tribe coalition was able to block two significant geothermal plants from being established. Although there is still the possibility that Calpine's leases will be extended, the company is guaranteed a long, difficult legal battle. Not only did the Pit River Tribe feel that the Modoc Plateau was worth the fight to protect it, the Advisory Council on Historic Preservation felt that parts of it were deserving of national historic recognition.

POLICY ARENA

The Department of the Interior identified the Medicine Lake area as having substantial alternative energy potential because of its remarkable geothermal capabilities. These resources were viewed as valuable to the DOI, even if this area were to be placed on the National Historic Register. Coincidentally, the DOI is not the only group who sees this area as highly valuable. It is worth noting that California has stringent environmental laws, most of which are stronger than their federal counterparts. The Calpine Corporation would have to complete all these requirements and adhere to these regulations in order to proceed with this project.

The Calpine Corporation has to deal with federal and local environmental regulations, on top of California's already stringent requirements. California has its own NEPA process, with tighter regulations and more reporting requirements than at the national level. Additionally, there are many strong conservation groups in California, arguably stronger than in many other states. It would not be a stretch to say that California is more liberal than many other states, and there is likely to an enhanced feeling of conservation there due to this environment.

Key Stakeholders

Several Native American tribes have a long, rich history of interacting with this area. Only one small coalition however, the Pit River Tribe coalition, felt this area is worth the time and effort to keep in its original condition. It is possible that other tribes mentioned in the case study felt that the benefits they would receive, such as jobs, scholarships, and an increased standard of living, were well worth any potential risks to their religious ceremonies performed in this area; they also may have felt that their time would be better spent fighting other battles.. The Pit River Tribe obviously values this religious site more highly than the other tribes. The other tribes and the Calpine Corporation found areas in which they are able to cooperate, a strategic move on the part of Calpine. The Shasta Tribe seemed to believe that they can capture more benefits once the government recognizes their Tribe, and the Calpine Corporation can help get that recognition. The Klamath and Modoc Tribes felt that any efforts would not be worth the possible gains captured, as demonstrated by their complete lack of engagement.

Joining the Pit River Tribes are two main conservation oriented groups. The Native Coalition for Medicine Lake Highlands Defense is a Native American conservation group. Second is the Mount Shasta Bioregional Ecology Center, an entirely conservation oriented group. Each of these groups share the common interest of wanting to prevent both the Fourmile Hill and Telephone Flats from being developed by the Calpine Corporation. The Calpine Corporation, on the other hand, felt and continue to feel that it is worth their time and money to invest in developing this site. The energy available to their company, and thus future profits, are expected to be worth the fight. Additionally, the fairly recent deregulation of the energy markets means the Calpine was able to enter the market with few other competitors. As time goes on more companies may try to enter this market, changing the playing field for Calpine. The Calpine Corporation appears willing to take the actions necessary to gain access to what they claim is their leased area.

RULES IN USE

As with many of these cases, and most environmental cases overall, one of the main laws in question is NEPA. One important requirement of NEPA is the completion of an EIS report, and particularly important to this case is when in the lease process a new EIS report needs to be conducted and when older EIS reports are sufficient. In order to save time and money both the US Forest Service and the Calpine Company would prefer to not redo these. The Pit River Tribe interests would best be served, however, if a new report were issued. Not only would the reinvestigation process give the Pit River Tribe more time to assemble a case, but they could also look for more groups to partner with. During this entire process, the Calpine Corporation was working against the clock to renew leases. It was Calpine's best interest, and that of the Forest Service (who would have to perform the Environmental Impact Statement), to not have to repeat the process.

Once the NEPA process had been completed, the Calpine Corporation had problems with the National Historic Protection Act. Not only did they have to worry that their current lease in Fourmile Hill would be lost due to historical protection concerns, but they would have to watch for similar problems with future leases. Their concerns became real when the Fourmile Hill lease was placed under protective watch. After this, Calpine Corporation had to wait five years before they could do anything with the site. If this wasn't frustrating enough, the new lease they purchased was placed under protective watch as well; the hold on the Telephone Flats lease was indefinite. As the most viable option to seek redress, it is no surprise that Calpine Corporation quickly moved to sue the government over its purchase of now unusable land.

Once in court, The Pit River Tribe (and their associates) argued that the Calpine Company had violated NEPA because it did not complete a sufficient EIS. Although two EIS statements had been done, the court ruled that none of were sufficient and that by extending the lease improperly, the Forest Service had violated NEPA and the National Historic Protection Act. This decision is a good example of the uncertainty inherent in the NEPA process. Even agencies that have a long history of compiling these reports can put together reports that are not considered satisfactory. While the courts established that Calpine Corporation did in fact have the right to develop on that land, it was that right which made them liable for another EIS statement.

Because these areas were close to lands protected by the NHPA, they violated the standards established within that Act by continuing their lease.

The court even goes as far as to elevate the NHPA to the same status as NEPA. The suit also stated that the Company's activities threatened species protected under the Endangered Species Act: the greater sage grouse, peregrine falcons, prairie falcons, and bighorn sheep.

The Pit River Tribe also argued that the Calpine Company violated the National Forest Management Act, which calls for forests to be managed in a way that best serves the public interest. The NFMA is an interesting case to consider because the original intent of this law was to promote alternative energy sources. It is ironic that it is now being used to prevent the siting of a alternative energy plant. The final environmental charge brought against them was violating the Federal Land Policy and Management Act. Again, this Act was designed to promote multiple-use land policies, but yet again it is used against Companies attempting to do just that. Further along this ironic path is the fact that the Pit River Tribe was able use the Forest Service's own report to prove that they had been injured by their actions.

INFORMAL RULES

Although NEPA is considered a formal rule, the way that it was used in the Fourmile Hill case can be understood as being governed by informal rules. Because informal rules are those that are unspoken, we believe this occurred as NEPA was interpreted by a set of previously unknown rules. In this case, the judges interpreted the law to the best of their ability, but the guidelines they used required a personal interpretation; there was no standard, outlined case to follow.

OUTCOMES

There are still many options for this case. First, the Calpine Company could be given the easy route and only be forced to apply for another lease extension. Once a new EIS is filed, and if no problems are found with it, this lease could be granted. There are also several options for the Pit River Tribe to utilize. First, they could stall the EIS application and lease process. They could do this by making excessive comments so as to create noise around the EIS. They could, and probably will, take the EIS statement to court to prove that it either isn't sufficient, or that the science in the EIS statement is not up to legal standards.

The Calpine Company could decide that the Pit River Tribe has put up enough of a fight that it is not worth doing another EIS and risking going to court again. They could look into other less promising, but less controversial geothermal sites. The Calpine Company could petition federal agencies, above those that are already on their side, to completely remove the site from the National Historical Registry and thus remove that future NEPA-like roadblock.

The most beneficial outcome for Calpine would be a lease extension for a period of time long enough to allow for proper exploration, and maybe even site development; this would require at least a forty-year lease. This outcome could also impact the other tribes besides the Pit River Tribe. Some of these tribes could be counting on the jobs that would come from a geothermal plant, not the mention the prospect of scholarships.

The Pit River Tribe has the option of petitioning to get the Modoc Plateau listed as a National Historic Site. Or, they could go beyond the NHPA and petition the President to establish the entire area as a National Monument, under the Antiquities Act. The Pit River Tribe can, and most likely will, continue to take the Calpine Company to court with any forward step the Company takes.

FUTURE

The most sought after outcomes for each party in this story are conflicting. A geothermal plant would provide clean, alternative energy for many Californians, citizens who are no stranger to energy shortages. Installing this plant, however, could encroach on the rights of several Native American Tribes to practice their religion and access sites critical to their history. Obviously a plan that allows for a geothermal site to be installed, while promoting access for these Tribes, would be a good compromise, but if the lawsuit is any indication, the Pit River Tribe isn't interested in having any energy plant on the site.

7 Hydroelectric Energy

INTRODUCTION

Since it was first proposed, Glen Canyon Dam has been controversial and opponents continue to advocate for removing or "decommissioning" the dam because it floods the area's natural landscape and ecosystem. The issues surrounding dam decommissioning are among the most controversial questions for hydroelectric power generation and are the focus of our case study.

Discovered by John Wesley Powell in 1869, the Glen Canyon region is some of the most rugged and inaccessible territory in the American West, but also some of the most beautiful land in the region. The Colorado River Compact of 1922 effectively divided the Colorado River's waters evenly between the upper and lower basin states. Construction for a dam was authorized but a site was not determined. In 1946, the Bureau of Reclamation published a study that identified Glen Canyon as a desirable site with promising hydropower capabilities. This declaration spurred a wave of environmental backlash, particularly from the Sierra Club, and protests slowed construction of many proposed dams. Amid years of heated controversy, Glen Canyon Dam was completed in 1963.

Environmental pressure to remove the dam continues and in 1989, the Secretary of Interior requested an Environmental Impact Statement. The EIS resulted in a reduction of the dam's output, which pleased environmental groups but failed to pacify them. Debates continue through the present day about whether or not the dam should be 'decommissioned' or removed.

HISTORY

The technology surrounding hydropower has been in use for thousands of years, with advanced developments beginning to unfold in France during the latter part of the eighteenth century (Water Power Program: History of Hydropower, n.d.). Ideas regarding water fueled turbines spread to and were further developed within the U.S. leading the way to the first ever hydroelectric power plant, which was opened for operation in Appleton, Wisconsin in 1882 (Water Power Program: History of Hydropower, n.d.). It was not long after this that hydro technology developed to the point where 25% of all electricity generated within the United States came from hydropower (Water Power Program: History of Hydropower, n.d.). These plants were used to power electrical grids, assist in building of dams and canals, and to power irrigation projects as settlers moved west. Hydroelectricity has seen a decrease over the years, from supplying 40% of the nation's energy needs in the 1900's to about 6% today (Water power Program: History of Hydropower, n.d.). It does remain, however, the second most common energy source in the world (How Hydroelectric Energy Works, 2006). There are over 2,000 hydropower plants in the United States. The number would be undoubtedly larger if there were more suitable sites (Bonsor, n.d.).

On September 30, 1882 the world's first hydroelectric power plant began operation on the Fox River in Appleton Wisconsin. President of the Appleton Paper and Pulp Company, H.J. Rogers, had been influenced by the plans of Thomas Edison to create a station in New York to produce electricity (Century, T.E., n.d.) and gathered together a group of prominent businessmen to tap into the perceived benefits of electric generation. Edison's plans involved cultivating steam energy for electric production; however, Rogers and his colleagues saw the Fox River as a more feasible source of energy and consequently used a waterwheel which initially powered the plant, Roger's home and another separate building (The World's First Hydroelectric Power Plant Began Operation, n.d.).

The early days of the plant were exciting times for the community as electricity began to be dispersed throughout the city, but the success of the plant was far from immediate. As the public gathered on September 27, 1883 for the scheduled opening of what would later be the Vulcan Street Power Plant, the bulbs refused to light. In response to this malfunction Edward T. Ames, the installer of the generator, had to be called back from Chicago to use a trial and error method to correct the unknown problem before a successful start was made on the thirtieth (Vulcan Street Plant, n.d.). Following the opening, a lack of technological advancements made it difficult to regulate the amount of electricity surging through the lines, thereby creating a wide variance in the illumination of the bulbs; many had such high levels of electricity running through them that the bulbs burned out.

Over time, the systemic kinks were remedied and the plant expanded, thus generating more power for a broader range of consumers. The once small company, established by investors through a risky business scheme, eventually made a large contribution to the economic success in the area and was responsible for the provision of electricity to the first hotel in the west (Vulcan Street Plant, n.d.). It is interesting to note the advancements in the small plant; at one time it only lit three buildings and produced 12 ½ kilowatts. It now serves over 93,000 customers with a generating capacity exceeding 600,000 kilowatts (Vulcan Street Plant, n.d.). The land on which the Vulcan Street Plant once stood in 1883 now houses the successful Wisconsin Michigan Power Company (Vulcan Street Plant, n.d.).

Further advancement of hydroelectric power quickly spread throughout the United States and Canada. Within three years of the Vulcan Street Plant's opening, forty to fifty different sites were in the beginning stages of either construction or planning (Bureau of Reclamation: History, n.d.). Additionally, the first dam built specifically for the purpose of supplying energy was constructed in Austin Texas in 1889–1893 on the Colorado River (Bureau of Reclamation: History, n.d.). Like the Vulcan Street Plant, this dam had great support from local businessmen, who hoped that the revenue and increased power supplied to the area would result in an economic boom. The granite structure was completed in 1893 and did in fact contribute heavily to the local economy through increased land sales; yet, the power supplied by the dam proved to be irregular due to variation in river volume. At times, the energy produced was barely enough to power the city's lights and streetcar system (Hunt, n.d.).

The magnificent sixty-foot high and twelve-hundred-foot long structure was considered the largest of its kind and brought with it a wide range of tourists who came to enjoy the newly formed Lake McDonald. Unfortunately, the enjoyment of the structure was short-lived; on April 7, 1900, the pressure of collected sediment at the base of the dam, along with extremely high water levels, led to its collapse, resulting in the death of eight people (Hunt, n.d.). Some efforts were made to restore the dam, but lack of support and continued flooding left it in ruin until the Lower Colorado River Authority rebuilt the structure, creating Lake Austin in 1940. This dam remains in use today (Tom Miller Dam and Lake Austin, n.d.).

Technological advances in hydropower led to increased implementation and consequently brought about the realization that such facilities had both positive and negative impacts on their surrounding areas. While dams had been proven to aid local economies with reduced electric prices and increased revenue, the impacts on the environment (which in turn harmed economic interests) led to federal legislation early on in the emerging days of hydropower. One such piece of legislation was the Tennessee Valley Authority Act, which was signed into law on May 18, 1933 by President Franklin D. Roosevelt (TVA: From the New Deal to a New Century, n.d.).

The demand for action in the Tennessee River Valley increased in the 1930's as the area struggled more than others through the Great Depression. The land in the area had been over-farmed, electric costs were phenomenally high, and the Tennessee River proved to be quite hazardous for travelers; therefore, President Roosevelt requested Congress create "a corporation clothed with the power of government but possessed of the flexibility and initiative of a private enterprise" to address the many issues in the area (TVA: From the New Deal to a New Century, n.d.). The Tennessee Valley Authority Act was therefore instituted as a federally owned agency dedicated to ensure proper use of the Tennessee River and its tributaries, as well as overseeing the supply of affordable electricity. This goal was met building dams throughout the area and promoting improved agricultural methods (Tennessee Valley Authority Act).

Some positive effects of the Tennessee Valley Authority's (TVA) efforts quickly became apparent. Flooding was controlled, for instance, but the greatest impact of the time was the spread of affordable electricity. Over time, TVA expanded by building three nuclear plants, eleven coal-fired plants, one wind-power site, and twenty-nine different hydroelectric power plants. The TVA has become a national leader in generating capacity-ranking at number five- and providing electricity to nine million customers across seven southern states (TVA: About TVA).

As development occurred throughout the western states, the use of water storage became a growing concern. Farmers and settlers commonly diverted streams and rivers for water supply, yet these practices alone were not enough to support the growing community and efforts were made at the local and state levels to store water for future use. A lack of advanced engineering led to failure for many of these storage undertakings and many were economically unfeasible, which prompted the U.S. government to establish the United States Reclamation Service to aid the western arid states in conservation and distribution of water.

The current mission of the Bureau of Reclamation is to manage, develop, and protect water sources in an economic and environmentally sound manner (Bureau of Reclamation, Mission, n.d.). It was not an immediate goal of the Bureau to develop hydropower; but the intermingling of water conservation and hydropower is evident through the Bureau's fifty-eight hydroelectric power plants across the western United States (Bureau of Reclamation, Main Page, n.d.).

INNER WORKINGS

Water reserves in domestic sectors of the global community have made hydropower a highly desired source of energy as an alternative to economic independence on foreign energy. Additionally, the availability of water and construction resources within the United States created jobs in local economies where hydropower is implemented. In addition, it secures low electric costs through the guaranteed abundance of replenishing water.

The first hydroelectric power plant in Appleton, Wisconsin provided 12.5 kilowatts for two paper mills and a home to be lit. Hydropower plants today vary from several hundred kilowatts to several hundred megawatts. A few massive plants have the capacity of up to 10,000 megawatts and provide electricity to millions of people. The combined capacity of hydropower plants worldwide is 675,000 megawatts. Annually, they produce over 2.3 trillion kilowatt-hours of electricity. This is comparable to 3.6 billion barrels of oil (Green Trust, n.d.). Power generated from hydroelectric plants makes up 65% of all alternative energy within the United States (Why Hydro, n.d.); this percentage, however, represents only six percent of the total amount of electricity produced within the country (EIA's Energy in Brief, 2012)

CATEGORIES OF HYDROPOWER SOURCES

Hydropower sources are divided into separate categories dependent on the amount of power generated, but the most common practice is that of using a dam to span a riverbed. In creating a dam, the water is elevated and therefore able to be channeled down through a tunnel below water lever called a penstock. As the water gains momentum through the penstock, it reaches and rotates a turbine connected to an electric generator.

In order to have enough kinetic energy to turn a generator, each gallon of water must be moving about one hundred feet per second. By building a dam, engineers can have direct control over where the water flows and how fast it is moving. Dams increase the distance water has to fall before reaching the turbines, allowing gravity to quickly increase the water's speed.

The basic structure of a dam is as follows. First, a giant wall is erected, forcing millions of gallons of water to wait to be funneled through penstocks. Penstocks are small tunnels through the dam wall that carry water down, allowing gravity to act on it and accelerate its speed. This water is pushed past turbines that turn a generator, turning kinetic energy into electricity. The generated electricity is then sent away along transmission lines.

There are two basic hydropower turbines: impulse and reaction. Impulse turbines use jets of water at atmospheric pressure. The most common impulse turbines are the Pelton wheel, which basically is a hub with a series of cups attached, the Turgo and the Crossflow, which operate similar to a boat propeller. (Hydropower Basic: Introduction, n.d.). Reaction turbines function when filled with water and utilize the angular and linear momentum of flowing water. Common turbines of this type include the Francis and Kaplan turbines (Campbell, 2010).

The Kaplan turbine, composed of three to six blades, uses the pitch of the blades to determine its overall performance, similar to a boat propeller. The Francis turbine resembles a weather vane and is the largest of the turbines. Here, one runner is attached to at least nine vanes, or blades, which the water turns. A third turbine type is the Pelton turbine. Here buckets are attached in an axial fashion around a disk. These are often used in very high locations and are the smallest of the turbines (How Hydroelectric Energy Works, 2006).

Further distinction can be made to include the means by which each hydroelectric plant manipulates water for energy production. Diversion uses fast rivers and waterfalls to turn a generator. A small section of the river is diverted away and used to rotate a turbine. The hydroelectric plant at Niagara Falls uses this means of creating electricity. This option is not as effective at producing electricity as damming rivers and results in less power output.

A third option is pumped storage (How Hydroelectric Energy Works, 2006). Here water is pumped from a low reservoir to a higher one, during an off-peak time. Later, when electricity demand is high this water is released to the lower reservoir, allowing gravity to act upon it. Some energy is lost during this process, leaving it with about 80% efficiency. Additionally, power is required to move the water from the lower reservoir to the higher one (How Hydroelectric Energy Works, 2006).

The U.S. Department of Energy categorizes the many different types of hydroelectric generators into three different classifications: large, small, and micro plants (Water Power Program: Types of Hydropower Plants, n.d.). Each classification is dependent on the amount of electricity generated, rather than the means used to create the electricity. Definitions vary according to source, but generally a large hydropower plant creates at least 30 megawatts of electricity, small plants generate between 100 kilowatts and 30 megawatts, while micro powered plants produce less than 100 kilowatts (Water Power Program: Types of Hydropower Plants, n.d.).

CATEGORIZATION-LARGE HYDRO

In the United States, the majority of hydroelectric facilities are storage plants. These use dams to store water in order to offset water flow fluctuations throughout the seasons. Storage plants can have a water capacity to operate for several years and provide a consistent source of electricity. The volumes of water flow and elevation differences, known as "head," determine the amount of electricity produced. The produced power can be utilized locally or carried through transmission lines to industrial or metropolitan users (Campbell, 2010).

CATEGORIZATION-SMALL HYDRO

As noted above, the term "small hydro" refers to hydroelectricity systems that produce less than 1000 kW per hour. Small hydropower holds many advantages over large-scale hydropower. It combines "the advantages of hydropower with those of decentralized power generation without the disadvantages of large-scale installations" (Hydropower Basic: Introduction, n.d.). Small hydro does not have near the costs for energy distribution or to the environment. According to Jon Wellinghoff, the chairman of the Federal Energy Regulatory Commission, "Efforts to reduce carbon emissions and meet the growing number of state alternative energy standards are drawing increased attention to small hydropower project development" (FERC looks to ease development of small hydropower projects, 2010).

Most of the more traditional-large hydroelectric sites have been developed. This has opened the door to small hydro. Areas where small hydro is applicable include small industries, farms or households, and rural communities. These are all decentralized locations with small demand for power (Hydropower Basic: Introduction, n.d.). Generally, small hydro utilizes low head sites, possessing an elevation difference of less than five meters. These types of systems often rely on run-of-river hydropower facilities that utilize the natural flow of stream and rivers. These facilities do not require large dams or reservoirs but some do have a small dam that simply keeps the impounded waters within the river's banks. Therefore, a natural flow of water continues over the dam. Another example of run-of-river hydropower is diversion facilities that channel a portion of river waters in an area that has a natural sharp decline, such as a river fall, to be utilized for power creation. This water is taken shortly off its course through the penstock and then the water is returned to the river further downstream (Campbell, 2010).

In 2006, the Idaho National Laboratory assessed 100,000 sites, throughout the United States, for small hydro capability and determined that approximately 5,400 had potential. The U.S. Department of Energy approximated that if these projects were developed, they would result in a more than 50% increase in total hydroelectricity generation (Campbell, 2010; Feasibility Assessment of Water Energy Resources of the United States, n.d.). With this supposed increase of hydroelectric power plants, the dependence that the world has on conventional energy such as coal and oil could be decreased.

CATEGORIZATION-MICRO HYDRO

Micro hydro is defined as producing an output less than 100 kW and generally is used to provide power to small communities or rural industry in remote areas by utilizing water resources available at a specific site. These systems generally use run-of-river systems and therefore do not require a dam. Consequently, they are dependent on the flow available which often varies with the seasons. Run-of-river systems consist of an intake structure which channels water through a pipe or conduit down to the turbines before the water is set free downstream. The speed of the turbine is generally controlled by a generator or alternator. Generally, at least three feet of head is required with a water flow of approximately twenty gallons per minute to run the turbine efficiently (Campbell, 2010).

ECONOMIC GROWTH V. ENVIRONMENTAL HEALTH

The growth of the hydropower industry has the potential to create 1.4 million jobs within the United States by 2025 (Why Hydro, n.d.). Currently, only 2,400 out of the 80,000 dams in the U.S. with hydroelectric capabilities, indicating this claim (provided by the National Hydropower Association) may be plausible. Hydropower is valued, "due primarily to the fact that once a facility has been built it is one of the least expensive sources of electricity to operate." (Campbell, 2010). That is, unlike other mainstream electricity sources, hydroelectricity requires no fuel costs. Dam maintenance, however, is a significant ongoing cost. This cost may be mitigated by the fact that the dam does more than produce electricity. It provides water storage for flood control, irrigation, and recreational activities.

Environmental concerns surrounding the construction of hydroelectric power sources include threatening fish populations and ecological changes in waterways. Although there are minimal air emissions and water is a renewable resource, it is clear that hydroelectric production produces several negative environmental effects.

A hydropower plant can prove to be detrimental to fish, plant life or other animals frequenting the area. Hydropower facilities disrupt the ecology of an area through deforestation and the increased collection of water. As the water levels increase, plant life is drowned and the temperatures and oxygen levels within the water decrease. Similarly, decaying plant life potentially increases the amount of the potential greenhouse gases, such as methane.

In recent years, hydropower has faced controversial concerns regarding potential negative environmental impacts associated with this form of power. For example, China's Three Gorges Dam, completed in 2006, has received a lot of attention in the media and in the surrounding communities. Some remarked on China's use of alternative energy, while others focused on the dam's displacement of the local community, approximately 1.9 million people (How Hydroelectric Energy Works, 2006).

Another example of a hydroelectricity controversy is the four large dams built on the Snake River. These four dams are part of the Federal Columbia River Power System, which is the largest energy source in the Pacific Northwest, and the largest alternative electricity source in the nation (Power Benefits of the Lower Snake River Dams, 2009, p.1). Since the construction of these dams, the native salmon population has dwindled to near extinction. Consequently, there is a high-profile national campaign advocating decommissioning of the four lower Snake River dams. Among the advocates of removing the dams are the Lower Colombia River Basin tribes who are guaranteed salmon by treaty. They are demanding that government honor native fishing rights and defend the Endangered Species Act (International Rivers, n.d.). Another way that advocates have been trying to increase the salmon population is to buy and lease water rights to those who are implementing better irrigation practices. They are creating meanders and other natural refuge areas so that the younger salmon have a place to grow without other invasions. (Habitat Restoration Means Business, 2011, p. 1)

In general, many have advocated a switch to Low Impact Hydropower in an effort to mitigate damage to fish migration patterns and other habitats (How Hydroelectric Energy Works, 2006). Additionally, a switch to small hydroelectricity has received significant praise lately; since no dams are built, the damage done to the surrounding wildlife is minimized. These small projects are also often more suited for rural areas where there is a small demand for electricity (Hydropower and the World's Energy Future, 2000, p. 10).

On the other side of the debate, there are economic benefits to be considered, such as flood control, barge transportation, irrigation, and especially electricity. The lower Snake River dams are integrated into the 500-kilovolt transmission grid from western Montana to eastern Washington. These Snake River dams are critical to keeping the system reliable (Power Benefits of the Lower Snake River Dams, 2009, p. 3).

According to Bonneville Power, "power production from the lower Snake River dams saves 4.4 million metric tons of CO_2 from reaching the atmosphere each year" (Power Benefits of the Lower Snake River Dams, 2009, p. 2). Regarding possible green energy replacement, Bonneville Power estimated the following figures based on wholesale power prices from March 2008, "$444 million to $501 million a year if the dams were replaced with natural gas-fired generation and $759 million to $837 million a year if the dams were replaced with a combination of wind, natural gas and energy efficiency. These figures are net of the dam's annual $38 million operation and maintenance costs" (Power Benefits of the Lower Snake River Dams, 2009, p. 4). The simple fact is, hydro is a cheaper green energy than any of its competing energy sources and productions.

SITING CONDITIONS

In order for a hydroelectric project to be sited, a location with all necessary conditions must be found first. It is almost absolutely necessary for a river to have a strong enough head and flow to produce electricity. Head can be either 'static' or 'dynamic' which is a measure of the water pressure as it falls (Oregon State Government, 2012). Flow is the rate, at which the water is moving and is measured in cubic feet per second (cfs) (Oregon State Government, 2012). The speed of the flow directly relates to how much power that will be available from that river or stream.

To achieve head to generate power, the hydropower plant must be located where water flows from a high elevation to a lower elevation. Generally, this occurs in mountainous areas, thereby eliminating many of the states in the plains. The river should be flowing in as straight a line as possible. Bends in the river should be far enough away from the site to allow the water to reach its maximum speed (Civil Features Volume IV, n.d., p.1). There needs to be space enough around the river to construct a powerhouse, turbine, and maintenance and service areas (Civil Features Volume IV, n.d., p.1). Additionally, there should be access roads to the site, to allow construction vehicles access. This river should also be close to the demand site, transmission lines or an electricity grid (Stojmirovic & Chu, n.d., p. 4).

REGULATIONS

In addition to the difficulty of locating an appropriate site, certain laws prohibit hydroelectricity development on certain federal lands (FERC: Off-Limits Sites, 2010). Here are a few of the acts that have an impact on hydroelectric power and the creation of hydroelectric plants.

NATIONAL WILD AND SCENIC RIVERS ACT

This act was established to protect rivers that display "outstanding remarkable scenic, recreational, geologic, fish and wildlife, historic, cultural, and other similar values." In addition, it preserves rivers "in free-flowing conditions, and protected for the benefit and enjoyment of present and future generations." For a river to become part of the Wild and Scenic Rivers System it must be done through an act of Congress or by the request of the state governor (FERC: Off-Limits Sites, 2010).

A "wild river area" is considered free of impoundments and usually inaccessible, with necessary primitive watersheds or shorelines and waters that are unpolluted. "Scenic river areas" are also without impoundments, and have primitive watersheds. At times it is accessible by road (FERC: Off-Limits Sites, 2010).

According to the Act, a hydropower project cannot be located "on or affecting" an area designated as wild or scenic river. This includes development above and below the designated river which are, "or on any stream tributary thereto," which will "invade the area or unreasonably diminish the scenic, recreation, and fish and wildlife values present in the area [as of] October 2, 1968" (FERC: Off-Limits Sites, 2010).

THE NATIONAL WILDERNESS ACT

This act prohibits any "commercial enterprise, structure, or installation within any wilderness area." A "wilderness area" is defined as "where the earth and community of life are untrammeled by man, where man himself is a visitor who does not remain." These areas are designated through an act of congress and a license or exemption cannot be issued for a project located within an area designated as wilderness (FERC: Off-Limits Sites, 2010).

ENERGY POLICY ACT OF 1992-NATIONAL PARKS

All properties under the jurisdiction of the National Park Service are considered part of The National Park including: national parks, monuments, recreation areas, trails, and heritage areas. According to the Energy Policy Act of 1992, it is prohibited to issue "an original license or exemption for any hydroelectric project located within the boundaries of any unit in the National Park System that would have a direct impact on the lands within that unit" (FERC: Off-Limits Sites, 2010).

DAM DECOMMISSIONING

Before 1994, the FERC had no policy for dam decommissioning. Nothing was stated in the Federal Power Act of 1920 regarding dam removal because it was assumed that the continued operation of dams was in the public's best interest. An official decommissioning policy was developed in 1994, but has only been implemented in 1999, with the refusal of re-commissioning for the Edwards Dam in Maine. The main reason that this policy has not been used more is because the majority of small, privately-owned dams are under the jurisdiction of state environmental agencies not federal. Interestingly, according to EOS, "Over one-third of all states in the nation do not have any statutes regarding dam removal. For the most part, statutes that do exist rarely go further than including removal along with repair or alteration as possible actions subject to regulation." (Geophysical Union, 2003, pp. 29, 32–33). In contrast there are a few states (Wisconsin, Pennsylvania, Ohio, and Connecticut) that "have adopted operational policies to expedite the process of approval for various permits needed for dam removal, and these states lead the nation in the numbers of dams removed" (Geophysical Union, 2003).

CASE STUDY—GLEN CANYON DAM

In 1869, John Wesley Powell, a one armed civil war veteran, set out with a group of mountain men to discover the unexplored portion of the West, specifically parts of Colorado, Utah, Arizona, New Mexico and Nevada. Before Powell's daring voyage, this area was merely black abyss on the map. Powell's adventurous nature led him down the Colorado River and through some of earth's most beautiful and untamed natural wonders.

After many close calls with Mother Nature and her raging rivers, the tattered group found themselves surrounded by smooth warm sandstone the color of salmon. The men were enveloped by mysterious sun streaked natural wonders and contours. As the peaceful calm surrounded them, the men were revived from the treacherous encounters with Mother Nature they had experienced only days before (Reisner, 1993, pp. 24–29).

These rugged outdoorsmen searched for a name that could properly express the awe and serenity they felt. On August 3, 1869, Powell wrote, ". . .we have a curious ensemble of wonderful features—carved walls, royal arches, glens, alcove gulches, mounds, and monuments. From which of these features shall we select a name? We decide to call it Glen Canyon" (Glen Canyon Natural History Association, n.d.).

This beautiful canyon, which offered a brief refuge to weary explorers, would be unrecognizable to Powell and his men today. The many natural beauties they experienced are now covered by approximately 9 trillion gallons of water. But before we jump ahead too far, let us go back and explain how this drastic change came to pass (Glen Canyon Natural History Association, n.d.).

Prior to Powell's astonishing discoveries in the West, the Industrial Revolution hit America, bringing with it a drive for innovation and power. Mills and factories began to dot the country and technological progress was the goal. As a result of this Industrial Revolution, the concept of dam building arose during the late 1800s and early 1900s, representing large symbols of power by man harnessing Mother Nature. The main purpose of this push was to power the new mills popping up throughout the United States. In addition, it soon became apparent that hydroelectricity was a profitable venture and larger dams were built such as the Hoover Dam on the Colorado River (1936) and the Grand Coulee Dam on the Colombia River (1942) (Citizens' Environmental Assessment on the Decommissioning of Glen Canyon Dam, 2000).

E. C. LaRue, an adventurous hydrologist for the U.S. Geological Survey, recognized Glen Canyon's dam possibilities in 1916 and a follow-up survey was performed in 1921. This area was attractive for dam building due to its narrow opening between 400 -foot sandstone cliffs. Only a year later the Colorado River Compact of 1922 emerged. This agreement divided the river's waters evenly between the upper and lower basin states, and specified the construction of a large dam along the Colorado to regulate the river's waters, but no specific location was given. States such as California, Utah, and Arizona battled for the right to have the dam in their state. These states perceived that a dam would bring water security, energy, and irrigation to their area and therefore, was a prize to be won. Ultimately, California's Black Canyon won but Arizonans were not about to give up their battle for a dam (Rogers, 2006, p. 9).

For decades, many prominent individuals proposed the plugging of the Colorado River in Glen Canyon but nothing happened until the Bureau of

Reclamation sponsored a major publication known as the "blue book." This study was published in 1946 and identified 134 potential dam sites throughout the Colorado River Basin. Glen Canyon was among the options with an estimated cost of more than $100 million. Sights were set on three locations for major dams: Bridge Canyon, Echo Park, and Glen Canyon. Each represented desirable hydropower capacity and all were in natural locations deemed critical to environmentalists. Therefore, the stage was set for disagreement.

All three proposed dams fell under heavy fire. Citizens were becoming more aware of environmental impacts of dams. From 1965–1970 the Sierra Club reported nearly a quadrupling of its membership. Aside from lobbying groups, some citizens' concerns led to action, sparking protests of many major dam projects (Citizens' Environmental Assessment on the Decommissioning of Glen Canyon Dam, 2000).

Echo Park, also named by John Wesley Powell and home to beautiful Indian rock art, generated the most determined opposition due to the fact that it would have flooded portions of Dinosaur Monument, a little known national monument in Southeastern Utah (May the Peace of the Wilderness Be With You, n.d.). Fighting for the integrity of the National Park System, organizations, specifically the Sierra Club, worked tirelessly for six years. These conservationist groups won, and their victory is seen as the birth of the modern environmentalist movement.

This environmentalist victory soon turned bittersweet. The trade-off the conservationists made to save Echo Park was an agreement to allow the remaining two proposed Colorado dams to be constructed, including Glen Canyon Dam (Citizens' Environmental Assessment on the Decommissioning of Glen Canyon Dam, 2000). During this period, tourists flooded Glen Canyon to witness the doomed splendors within its walls. The canyon walls and river bottoms were vivid reminders of rich history such as Moki Indian remains, steps carved into the wall by the Dominiquez-Escalante party, inscriptions by Powell and his men and wagon ruts from early Mormon pioneers. Reclamation officials attempted to document the history within its walls before it was all covered with water. Doing this was, however, secondary to the primary task at hand—building a dam (Rogers, 2006, p. 8).

Many efforts were made to halt the dam's construction. One of the most memorable was a full -page ad in the New York Times asking, "Should we also flood the Sistine Chapel so tourists can get nearer the ceiling?" (Citizens' Environmental Assessment on the Decommissioning of Glen Canyon Dam, 2000). Ultimately, these campaigns were unsuccessful and the Glen Canyon Dam was completed in 1963, and the trapped waters began to cover the historical and natural gems. At the time, David Brower, Sierra Club's executive director, called its construction "America's most regretted environmental mistake" (Citizens' Environmental Assessment on the Decommissioning of Glen Canyon Dam, 2000).

Yet, many saw it as the beginning of a new period in the West. Commissioner Dominy gave a speech in April 1965, in which he celebrated the newly formed reservoir "for its many functions—drinking water, electric power, tax revenue . . . (and) use as a recreational haven" (Rogers, 2006, p. 32). The new dam became a vital link to providing electricity, water, flood control, and recreation to the West and her people. At 710 feet, the dam held back an amazing 26,215,000 acre-feet of water and continues to do so today, creating the 180-mile-long reservoir known as Lake Powell (Glen Canyon Natural History Association, n.d.). The main purpose of the dam is to store water so that the water and power can be delivered to cities and towns in the Southwest such as Las Vegas, Los Angeles, and San Diego (Rogers, 2006, p.31).

Recreation has also been a highly praised advantage of Glen Canyon Dam. In 1965, approximately 500,000 vacation -seeking individuals visited the lake. Consequently, the Department of Interior built marinas and other internal improvements costing millions. In addition, the need for sewage, trash and water pollution management has been ever increasing (Rogers, 2006, p.32). The lake continues to be popular for recreation with nearly 3 million annual visitors as many as 200,000 on the Fourth of July.

There are serious ecological impacts of the dam. "The presence of any dam presents challenges to the river system and its habitat. Whereas trout could not survive in the free-flowing muddy Colorado, for instance, they flourished in the reservoir" (Rogers, 2006, p.32). Shortly after the dam was completed, twelve million trout and five million bass were dumped into the newly formed Lake Powell. Trout thrive in the cold waters of Lake Powell; however, the cold clear waters released by the dam do not provide the warm waters necessary for most native fish, such as the scale-less Humpback Chub and Colorado Pikeminnow, the largest American minnow measuring up to six feet and eighty pounds in size (San Juan River Basin, 2007). The public became aware of the dam's impact on the river system and in the 1980s there arose expressed concern over "beaches, endangered species, ecosystem, fish, power costs, power production, sediment, water conservation, rafting/boating, air quality, and the Grand Canyon wilderness" (Rogers, 2006, p. 33).

President Reagan's Secretary of the Interior James Watt responded to public, environmental, and tribal pressure by initiating the Glen Canyon Environmental Studies (GCES). These were a series of scientific studies on the Colorado River to evaluate dam operations and its environmental impacts on Glen Canyon and the Grand Canyon downstream. The studies concluded that heavy release of clear cold water from the dam did negatively affect the downstream ecology. However, environmentalists were not satisfied because the GCES did not require any action as a result of its findings.

Public pressure built and in 1989, the Republican Secretary of Interior Manual Lujan announced that an EIS would be produced. This became known as GCES II. These studies were highly debated and complex. The EIS was released in 1995 and "provided methods of preserving cultural artifacts, flood control, (and) habitat" (Rogers, 2006, p. 34). Commissioner Daniel P. Beard commented that, "the EIS was the demarcation line between the old way we have treated the river, and the way of the future" (Rogers, 2006, p .34).

The main change that the EIS caused in the daily function of Glen Canyon Dam was to decrease the amount of output. While in general, most environmentalists agree that this is a positive change on the river's ecology, many argue against the mere existence of the dam. In 1981, a radical environmental organization, Earth First!, made a dramatic demonstration by unrolling a three-hundred foot long plastic "crack" along the front of the dam. Since then, other organizations have joined in the fight to drain Lake Powell and restore the free-flow of the Colorado River (Rogers, 2006, p. 35).

Some claim,"It's not a question of if a dam will need to be removed; it's only a question of when" (Citizens' Environmental Assessment on the Decommissioning of Glen Canyon Dam, 2000). Many advocacy groups have voiced concern about the dam's structural security. The Glen Canyon Institute estimates that the Glen Canyon Dam has a useful lifespan of approximately one hundred more years. A comprehensive facility review was performed in 1998 and concluded that, "despite some cracking, the foundation and structure are in excellent form" (Rogers, 2006, p. 35). The Glen Canyon Institute claims that the dam is filling up with sediment, which places pressure on the massive cement wall and creates a growing threat of extreme ecological damage when the reservoir waters are finally released (Citizens' Environmental Assessment on the Decommissioning of Glen Canyon Dam, 2000).

The general popularity of dam removal, also known as decommissioning, is gaining public support. Over the last forty years, more than 460 dams have been removed. "At the heart of the River Restoration Movement is the Glen Canyon Dam and the movement to restore a healthy Colorado River. The loss of Glen Canyon was a turning point in the birth of the modern Environmental Movement and has been mourned since the dam's construction" (Citizens' Environmental Assessment on the Decommissioning of Glen Canyon Dam, 2000). In the forward to the book titled, *The Place No One Knew* by Eliot Porter, Brower lamented over the loss of Glen Canyon. "Glen Canyon died in 1963 and I was partly responsible for its needless death. So were you. Neither you nor I, nor anyone else, knew it well enough to insist that at all costs it should endure" (Rogers, 2006, p. 27).

The debate surrounding the legitimacy of Glen Canyon Dam continues today. Main supporters of the dam include the Bureau of Reclamation, Western Area Power Administration, Colorado River Energy Distributors Association (CREDA), and surprisingly the Utah chapter of the Sierra Club. According to Ann Wechshler with the Sierra Club in Utah, the chapter has "an obligation to keep the memory of Glen Canyon alive and never to allow that kind of boondoggle to happen again. But Glen Canyon is gone. Maybe in the future we can talk about dismantling dams on the Colorado. Right now, there is not the local support for doing that" (Current Controversy: Draining Lake Powell, 2002).

Those in opposition to the dam include the national Sierra Club, Glen Canyon Institute and Earth Island Institute. These organizations are active in informing the public of the biological, physical, and economic concerns surrounding the dam and are actively seeking the decommissioning of the dam. In December of 2000, the Glen Canyon Institute released their initial Citizens' Environmental Assessment (CEA). This study examined the opportunities, costs, and environmental impacts that would result if Glen Canyon were restored. In addition, these organizations fund numerous studies focused on better understanding the overall effects of Glen Canyon with and without the dam (Citizens' Environmental Assessment on the Decommissioning of Glen Canyon Dam, 2000).

FUTURE OF HYDRO

As scientists try to map out the future of hydropower, some are forced to grapple with the effects of climate change as an increasing number of rivers are drying out. Dryer conditions are expected to greatly affect South Africa, Afghanistan, Brazil, Tajikistan, and Venezuela (Corley, 2010). Other areas are expected to see a boost in hydroelectric potential with the changing climate, namely: Eastern Africa, Southeast Asia, and Northern Europe (ibid). Corley, 2010 Researchers are looking at ways to increase the efficiencies of turbines in order to quickly adapt to the decreasing water levels.

Essentially, however, water is an inexhaustible source because once the water is used to produce electricity, it is usually returned to its original river or reservoir. Because of this, hydropower has great potential now and also into the future. Only 20 percent of hydroelectric potential has been developed within the United States; tapping into this potential however, is often hindered by unsuitable terrains and large distances from needing communities (Hydro Power Plants, n.d.) In addition, future projects do not necessarily require new locations or dams. Only 2,400 out of 80,000 dams in the U.S. currently produce electricity from hydroelectric power plants. Many of these existing dams could have advanced technologies installed to produce energy and increase efficiency. A study performed by the U.S. Department of Energy (DOE) calculated that it would cost approximately $1,600 per kilowatt to add turbines to dams that currently lack electricity capabilities (Campbell, 2010). With this data and the amount of potential future locations for hydroelectric power plants, the cost the retrofit the existing dams would be able to pay for themselves in just a short time. Further these attempts are complicated by the push for additional dam decommissioning by local and national environmental groups.

In addition, the DOE's Idaho National Laboratory did a study on the potential of developing small and low head hydropower sources in the U.S. Specific criteria were set to determine the feasibility of a site. "These criteria assumed that a dam would not be required, that sites were close to towns or electricity lines and roads, and penstock lengths were based upon those at existing low power and small hydroelectric plants" (Campbell, 2010). Approximately 5,400 of the 100,000 sites considered were identified as having potential for small hydro projects. For this study, the Idaho National Laboratory defined small as "providing between 1MW and 30 MW of annual mean power" (Campbell, 2010). According to DOE estimates, if these projects were developed they could increase total hydropower generation by fifty percent.

8   Biofuel Energy

INTRODUCTION

Environmental factors and the ever-developing scarcity of petroleum throughout the world have, over the last forty years, led to an increasing governmental policy focus on alternative fuels. Today the two most common of these petroleum alternatives, ethanol, and biodiesel, see wide (and accelerating) use across the world. This massive expansion in use is due largely to national policies, which have led to large modifications to their inherent supply and demand. In this chapter we will examine the history, related policies, economic viability, and environmental impacts of these alternatives to traditional fossil fuel based energy.

The vast majority of worldwide transportation today is accomplished by the use of internal combustion engines, which are typically fueled by petroleum-based combustibles such as gasoline and traditional diesel. The two standard types of engines, gasoline and diesel, are generally referred to by the subtype of petroleum fuel they use for combustion. The two engines differ in how they cause combustion, and the properties of the fuel each uses. Gasoline fueled engines use a spark to initiate internal combustion, while diesel fueled engines rely solely on heat generated by compression of the fuel to cause ignition (Brian, n.d.). Given the different approaches, the appropriate fuel for the engines must have characteristics specific to each particular application, and they are not interchangeable. Historically, petroleum based fuel has supplied the overwhelming majority of the demand worldwide, but alternative fuels for internal combustion engines have existed as long as the engines themselves.

Ethanol can be used as a replacement (or supplemental additive) for petroleum-based gasoline (petrol gas), while biodiesel is used in place of (or as an additive to) petroleum based diesel fuel (petrol diesel). The characteristics and sources of these biofuels have profound impacts on both vehicle fuel economy, and various chemical emissions released through combustion. Although ethanol and biodiesel generate less power per volume of fuel consumed than their petroleum based counterparts, studies have found a reduction in most greenhouse gasses traditionally found in tailpipe emissions. A perceived overall reduction in undesirable emissions has been the primary motivating factor in influencing a broad spectrum of federal policies and regulations enacted in the United States. This perception, however, has been directly called into question by several scientific studies, which found that the actual carbon impact of both biofuels will vary dramatically according to the source of the fuel, and the method of accounting used in analyzing impacts. Other potential issues addressed in this chapter include the impact of biofuels on staple food crop prices and availability, biodiversity (in certain environments), impacts on water consumption, and unaccounted for impacts of greenhouse gas (GHG) emissions related to pesticide and fertilizer consumption.

HISTORY-BIODIESEL

Biodiesel is a new take on the petroleum diesel that many have come to know since the very late nineteenth century. Diesel engines are known as such for their inventor, Rudolf Diesel, whose engine came about to replace the extremely inefficient steam engines which were in use in the 1890's (de Paula, 2011); these diesel engines were initially fueled by peanut oil, the first biodiesel. Biodiesel was quickly replaced with petroleum-based diesel as the primary fuel as petroleum diesel has been, and continues to be, more cost effective to produce. Biodiesel may not be in the spotlight anymore, but it certainly has not lost too much ground.

The diesel engine replaced steam engines in the early twentieth century due to its higher thermodynamic efficiency, and remains in use today in a broad variety of applications. Diesel engines are preferable to gasoline engines in certain applications due to higher torque generated for a given engine displacement, higher efficiency per gallon of fuel consumed, longer engine life, and lower CO_2 emissions (Diesel—Overview, 2011). The lower emissions of CO_2 are offset, however, by higher particulate matter and NOx emissions compared to gasoline-fueled engines. With the advent and phase-in of low sulfur diesel fuel, emissions of SO_2, which can cause acid rain, have been dramatically reduced. This reduction in SO_2 is, to a degree, mitigated by higher emissions of NOx, which can also cause acid rain.

In the United States, diesel engines have been most commonly used in large marine applications, heavy trucks, and trains. Although relatively uncommon in US passenger cars, diesel engines comprise nearly half of all new vehicle sales in Europe (Diesel—Overview, 2011). There are a variety of reasons for the lower use in the US. Cost conditions of both the diesel engine itself, commonly $1,500 more than a gasoline engine option (Diesel—Overview, 2011), and the higher price of diesel (as compared to gasoline) have strongly influenced the adoption of diesel engines in passenger vehicles in the United States. Since 2004 the US price of diesel fuel has generally been higher than the price of gasoline, due in part to world demand conditions, and part to the higher ($0.06 per gallon higher than gasoline) federal tax on diesel fuel (Webster, 2009).

Historically, the primary fuel used for diesel engines has been petroleum-based diesel due to the lower cost of production; however, as petroleum has become scarcer and demand has increased, the price of petrol diesel has rapidly approached that of biodiesel. A variety of laws and statutes over the course of the last forty years has also influenced the viability of biodiesel. Proponents of biodiesel regularly cite the advantages of biodiesel such as significantly reduced emissions, its ability to produce large quantities domestically, and its overall renewability. Opponents' arguments, however, critique the recoupment time of the initial release of carbon from clearing land for planting crops used to supply the bases for biodiesel, the increased cost of food caused by demand for source crops, and some efficiency issues.

Biodiesel is derived from various organic sources. These sources include: soybean oil, peanut oil, cottonseed oil, canola oil, recycled cooking greases or oils, and animal fats (beef tallow, pork lard) (Maryland Energy Administration, n.d.). Expedito Parente developed several techniques for the refinement of biodiesel in 1977, which enabled for more efficient, modern scaled production. Biodiesel is typically produced by a chemical reaction between methanol and the source using a catalyst (National Biodiesel Board, 2011). This process, called transesterification, separates glycerin from the source leaving methyl esters, the chemical name of biodiesel (National Biodiesel Board, 2011). Methyl esters are typically mixed with conventional petroleum based diesel fuel in a variety of blends, but can also be used without blending. The byproducts of the process are generally glycerin and solid waste from the source material (which can often be used as feed for livestock), both of which are marketable goods.

While vegetable oil from soybeans is the most prevalent source of biodiesel in the US, any plant that produces a sufficient quantity of lipids can be used as a source for biodiesel (though some are much more efficiently converted than others). Climatological conditions in the US are what make production of soybeans the most common source; however, several other sources are in development, including rapeseed oil (occasionally referred to as canola oil.) Animal fats are quite easily converted to a biofuel, and the use of fat waste products is quite common in small-scale production of biofuel. Research into lipid producing algae has also shown large amounts of promise as another source of biodiesel. Outside of the US, the most common source crop for biodiesel is palm oil, which is commonly produced in tropical climates.

The uses of biodiesel are varied. The highest quality amount you can obtain is the purest, which is labeled as B100. The higher the number the more biodiesel is included in the fuel. Other blends work equally as well. One of the most common is B20, a blend of 20% biodiesel and 80% diesel fuel. Anything above the B20 blend may require alterations to the diesel engine to be able to run more fluidly and properly. Other lower blends such as B5 and B2 are used more publicly than the others. The B20 blend is used widely in many different fleet vehicles, with the largest and most well know being of the United States Postal Service. In 2005 the US Postal Service consumed over one million gallons of B20 biodiesel with their various service vehicles, more than any other alternative fuel provided, besides the conventional gasoline and diesel (Buchholz et al., 2005, pp. 4–5). The primary cities that have these vehicles are New York, Detroit, St. Lewis, and Orlando.

Despite subsidies and protectionist policies, biodiesel remains more expensive than petroleum based diesel. As of July 2011, the nationwide average price of biodiesel (B20) was $3.67 per gallon and with pure biodiesel being priced at $4.13, compared to the petrol diesel price of $3.54 per gallon (Clean Cities Alternative Fuel Price Report, 2011).

According to the National Renewable Energy Laboratory (NREL), the cited advantages of biodiesel are: renewability, energy efficiency, reduction of global warming gases and emissions, non toxicity, and biodegradability (National Renewable Energy Laboratories, n.d.). With the entire lifecycle-CO_2 balance accounted for, use of biodiesel can reduce greenhouse gases by up to 41% if produced from crops harvested from existing productive fields (Biodiesel, 2011). This figure incorporates the consumption of CO_2 by the source plants, which reduces the net CO_2 impact of the use of biodiesel. A separate study found that the use of B100 reduces CO_2 emissions by more than 75% as compared to petrol diesel with the effects of source plants incorporated (Alternative Fuels and Advanced Vehicles Data Center: Biodiesel Benefits, n.d.). Neither of these studies took into account the carbon-debt of land clearing, which will be addressed later in the chapter. This reduction is due to the higher oxygen content (11% as compared to petrol diesel content of roughly 0%) of biodiesel fuel that combusts more completely, resulting in lower hydrocarbon and carbon monoxide emissions from most engines (Biodiesel, 2011). In addition, biodiesel generally contains no sulfur, which reduces the total tailpipe emission of SO_2 in proportion to the mixture used. This decrease in emissions comes at a cost. Use of B100 fuel reduces available torque by 7% on average (as compared to petro-diesel), whereas use of B20 fuel typically reduces torque by less than 2% (Sharp, n.d.). The loss of torque is due to the lower volumetric energy content of biodiesel as compared to petrol diesel.

The American Society for Testing and Materials (ASTM) has established standards that regulate the content of biodiesel (Proficiency Testing, 2012). Biodiesel in compliance with standards outlined in ASTM D6751 can be used in standard diesel vehicles without modification (National Biodiesel Board, 2011). Use of biodiesel in some vehicles, however, voids the warranties offered by manufacturers if the approved blend is exceeded. Here is a short list:

Acura: no current US diesel models
Audi: up to B5 blends
BMW: up to B5
Chrysler: B20 approved for Dodge Ram for government, military, and commercial fleets, B5 for all other
Cummins: B20 approved
Ford: B20 approved for model year 2011 and forward, B5 for 2010 and prior
General Motors: B20 approved for all 2011 and forward, B5 for all other GM diesel vehicles
Honda: no current US diesel models
Hyundai: no current US diesel models
Isuzu: B5 approved (in process of researching B20 for approval)
John Deere: B20 approved (B5 preferred)
Mack: B5 approved
Mercedes Benz: B5 approved
New Holland: B100 approved
Nissan: TBA (as of 2009)
Toyota: TBA
Volkswagen: B5 approved
Volvo: B5 approved

Biodiesel has an increased wearing effect on some engine and peripheral components, softening and degrading some rubber compounds over time. Blends below B20 are less likely to degrade rubber, and most manufacturers approve blends no higher than B5. Biodiesel has a somewhat higher cloud and gel point than petrol diesel, which can affect performance at low temperatures depending on the blend of biodiesel in the fuel. Biodiesel may also be used for home heating oil, with B5 meeting the ASTM D396 standards for home use (Biodiesel, 2011).

In contrast to petroleum-based diesel, which is highly toxic, biodiesel has roughly the toxicity of table salt (Environmental and Safety Information, 2012). The skin irritation resulting from accidental contact is generally no more irritating than strong hand soap. If spilled in water, biodiesel degrades by up to 88% within twenty-eight days (Environmental and Safety Information, 2012). Furthermore, the flashpoint (temperature of ignitable vapor formation) for biodiesel is 150°C, compared to 52°C for petrol diesel. Given these factors, from a storage and safety standpoint, biodiesel has several advantages to petrol diesel.

HISTORY-ETHANOL

Ethanol was first used as a lighting fuel during the 1850's. During the Civil War, there was a tax levied upon ethanol to help fund the war effort. The price of ethanol was increased as a result of the tax, which led to it being unable to compete with other forms of fuel, which led to a downturn in its use and production. These levels of production did not increase until the liquor tax was repealed in the year 1906. During the years of prohibition, pure ethanol was banned due to its status as a liquor. It also was not mixed with petroleum at this time, and was not used as fuel until prohibition was lifted in 1933. Ethanol wasn't widely used from this point up until the 1980's when the United States began to use ethanol as a replacement for the lead found in gasoline. In Brazil, however, there was a nationwide pro-alcohol program began to finance the government and enter into the global fuel economy. This program was effective at reducing Brazilian dependency upon fuel imports, which contributed to the rapid growth of their economy during that time (Ethanol Facts and History, n.d.).

Ethyl alcohol (or ethanol) is a alternative fuel produced by milling grains (in most cases corn) into meal, which is then combined with water and enzymes to convert the starches to dextrose (How Ethanol is Made, n.d.). This is then combined with yeast, which digests the sugars, producing ethanol and carbon dioxide in a process similar to beer production (How Ethanol is Made, n.d.). The resulting fluid is then distilled and filtered to obtain pure ethanol, the same chemical compound found in many alcoholic beverages (Alternative Fuels and Advanced Vehicles Data Center: Ethanol, n.d.). The resulting product is combined with gasoline in various mixtures and sold as fuel for vehicles. The remaining solids and liquids are generally sold as animal feed, while the CO_2 produced can be captured and marketed for use in carbonated beverages or dry ice (Alternative Fuels and Advanced Vehicles Data Center: Ethanol, n.d.).

Some sources, such as the Renewable Fuels Association (RFA), claim reductions of up to 29% in CO_2 emissions (as compared to E0 fuel), and potentially higher reductions incorporating the CO_2 recycle effect of the feed stock (How Ethanol is Made, n.d.). In other studies, the effects of ethanol on emissions over the complete fuel-cycle have been shown to vary widely with the process, and fuel used in the production of ethanol, from a 3% increase in greenhouse gas emissions when coal is used, to a 52% reduction if wood chips are used (Wang, Wu, & Huo, 2007). The impact on greenhouse gas emissions comes from the use of biological feedstock in production that consumes CO_2, and also from the higher oxygen content of ethanol, which results in a more complete burn and lower tailpipe emissions. The feedstock is typically corn, though several other sources are in development including technologies that produce ethanol from the cellulose in a variety of plant sources. Measures of the CO_2 impact typically cover from growth of the product, through production processes, and out the tailpipe of a vehicle (cradle to grave). The initial impact of converting land to production is not typically included. As in the case of biodiesel, clearing previously unutilized land to production can significantly modify the figure for carbon impact. This is commonly referred to as "carbon-debt," and will be discussed later on in this chapter.

Commercial fuel contains ethanol, and like biodiesel, is typically listed by the letter E, followed with the percentage mix, which corresponds to the proportion of ethanol in the fuel compared to the petroleum content. E10 denotes a fuel mix containing ten percent ethanol, and ninety percent regular petroleum gasoline. E10 is the most common form, and is found at many fuel stations throughout the United States.

Pure ethanol has approximately two thirds the energy content of unblended petroleum based gasoline, resulting in significantly reduced fuel efficiency (Impact of Ethanol Blending, 2008). A study by the National Renewable Energy Laboratory found a loss of 3.88 miles per gallon from E0 to E10, and 7.72 from E0 to E20 (Impact of Ethanol Blending, 2008). Use of E85 in flexible fuel vehicles (FFVs) typically results in a 25–30% reduction in miles per gallon (Fuel Economy, 2012). Even an ideal process used to produce ethanol could result in an increase in total tailpipe emissions of CO_2, taking ethanol's reduced fuel efficiency into account (as compared to pure gasoline).

Biodiesel is available in many areas of the United States, with a very high concentration in Virginia, North and South Carolinas, and Tennessee (BioFuels Atlas, 2011). Annual production of biodiesel in the US reached nearly 700 million gallons in 2008, which was up from 25 million in 2004, with over 1,900 distributors existing nationwide (U.S. Department of Agriculture Factsheet, 2010). This expansion of production is, in part, a response to federally mandated volumetric requirements (of both vehicles and fuel) and tax incentives. A voluntary quality control program developed by biodiesel producers and distributors called BQ-9000 certification has accelerated acceptance of the product by both customers and vehicle manufacturers (BQ-9000, 2011). This certification is granted to producers who meet all ASTM standards regarding biofuel quality, in addition to further rigorous quality assurance requirements.

Ethanol is typically found as a fuel additive in nearly every gallon of gasoline sold. With the increased prevalence of flexible fuel vehicles however, higher percentage mixtures are becoming widely available. As of January 31, 2012, the Alternative Fuels & Advanced Vehicles Data Center (AFDC) reported 2512 stations in the US carrying E85 fuel (Fueling Station Locator, n.d.). In certain regions of the United States it is possible to find gas stations that offer fuel, which is based mainly upon ethanol such as E85. The large majority of these stations are located in the Midwestern United States. Since the United States is one of the largest worldwide producers of corn, it is feasible to expect greater usage of corn for ethanol as the popularity of ethanol increases. In 1992 the Energy Policy Act was passed which defined ethanol blends containing at minimum 85% ethanol as an alternative fuel. Those who provide ethanol are given tax breaks to encourage the development of an ethanol car market within the United States. By the year 2005, there were over four million flexible fuel vehicles capable of running on E85 fuel. Currently, there are over 400 gas stations that contain E85 fuels. (Ethanol History-From Alcohol to Car Fuel, n.d.).

Since biodiesel is not as commonly available as ethanol, and neither is as widely available as petroleum fuels, many resources exist to assist consumers in finding fueling stations. As of January 31, 2012 AFDC lists 637 stations nationwide carrying B20, primarily in the east and Midwest. The US Department of Energy (DOE) sponsored the creation of the Alternative Fuel Station Locator, an interactive program that locates alternative fuels, and can be used on either computers or handheld devices (Fueling Station Locator, n.d.). This application locates nearby alternative fuel stations by entering a zip code, and filters by the type desired, providing addresses, phone numbers, and accessibility for available stations. As of December, 2011, the DOE reported 285 publicly available, and 348 private access stations carrying blends of B20 and above throughout the United States. The vast majority of these were listed as "private-government only." At many stations, the biodiesel is blended with petrol diesel to order, providing some flexibility in blends available to the consumer (Fueling Station Locator, n.d.).

REGULATIONS

Key legislation influencing the adoption of biodiesel (and ethanol, among other biofuels) include the Clean Air Act of 1970 (CAA70), the Alternative Motor Fuels Act of 1988 (AMFA88), the Energy Policy Act of 1992 (EPAct), which has been amended by the Energy Conservation and Reauthorization Act of 1998, and the Energy Policy Act of 2005, the Energy Independence and Security Act of 2007 (EISA), the Energy Improvement and Extension Act of 2008, the American Recovery and Reinvestment Act of 2009, and the Tax Relief, Unemployment Insurance Reauthorization and Job Creation Act of 2010 (Alternative Fuels and Advanced Vehicles Data Center: Key Federal Legislation, n.d.).

Several events have contributed to the degree to which ethanol is produced and used in the United States. Beginning with the Energy Tax Act of 1978, tax exemptions of $.40 per gallon or more have been available to ethanol related industries (Ethanol Overview: Industry Growth and Federal Programs, 2011). The policy of promoting ethanol use expanded to include an import duty for fuel ethanol with the Omnibus Reconciliation Act of 1980, an increase in tax credit with the Deficit Reduction Act of 1984 (to $.60 per gallon), and a tax credit of $.10 per gallon for producers with the Omnibus Budget Reconciliation Act of 1990 (Ethanol Overview: Industry Growth and Federal Programs, 2011). A broad variety of grant and loan programs exist to aid development and production of biofuels in the US including: the USDA Bioenergy Program, the Biomass Research and Development Act of 2000, and the Healthy Forests Restoration Act of 2003. The Bioenergy Program makes payments from the USDA's Commodity Credit Corporation (CCC) to biofuel producers based on quantity increases with a goal of encouraging purchases of biological sources for biofuels (Ethanol Overview: Industry Growth and Federal Programs, 2011), while the Biomass Research and Development Act of 2000 created programs intended to increase research and development in biofuels (Ethanol Overview: Industry Growth and Federal Programs, 2011). The Healthy Forests Restoration Act of 2003 amended the Biomass Act of 2000, expanding grant and loan programs greatly for overall biomass (Ethanol Overview: Industry Growth and Federal Programs, 2011).

The Clean Air Act of 1970 outlines the responsibilities of the Environmental Protection Agency in protecting and improving air quality. Under CAA70, the developments of regulations with the intent of limiting emissions from various sources were authorized, including the establishment of the National Ambient Air Quality Standards (NAAQS) (Overview- The Clean Air Act Amendments of 1990, n.d.). In 1990 the CAA was amended, providing additional regulatory authority to the Environmental Protection Agency (EPA), and establishing additional fuel quality controls and pollution standards (Overview- The Clean Air Act Amendments of 1990, n.d.). The emissions standards emplaced by the CAA70 (amended in 1990) created a tax incentive system that has led to the use of a variety of cleaner burning, alternative fuels such as ethanol and biodiesel (Overview- The Clean Air Act Amendments of 1990, n.d.).

The AMFA88 created manufacturer incentives through the Corporate Average Fuel Economy (CAFE) program in the form of tax credits for production of motor vehicles capable of operating on alternative fuels (Alternative Fuels and Advanced Vehicles Data Center: Key Federal Legislation, n.d.). AMFA88 also required the creation of a data resource and education center for alternative fuels, which led to the formation of the Alternative Fuels and Advanced Vehicles Data Center (AFDC) in 1991 at the National Renewable Energy Laboratory (NREL). The AFDC is a Department of Energy funded research and information database focused on alternative fuels and advanced transportation technologies.

Biodiesel was recognized as an alternative fuel under the Energy Policy Act (as amended in 1996 and 1998), which required that no less than 75% of new vehicle purchases by federal and state fleets be alternative fuel vehicles (Frequently Asked Questions—EERE. 2009 p.2). This alternative fuel fleet requirement has led to a significant proportion of the demand for biofuels. An excellent example is the fleet of the United States Postal Service, which consumed over one million gallons of B20 in 2005. In the disassembly of Postal Service vehicles that ran in service from the 2000 to 2004, with six hundred thousand miles on each vehicle, there were some issues that might be blamed on the use of biodiesel B20 blend. The Mack tractors that were used exhibited higher frequency of fuel filter and injector nozzle replacements over the years. Biological contaminants may have caused the filter plugging. There was also a sludge buildup that was noted around the rocker assemblies in the Mack B20 engines. The sludge contained high levels of sodium, possibly caused by accumulation of soaps in the engine oil from out-of-specification biodiesel (Buchholz et al., 2005, p. 6). These issues do not necessarily mean that the fuel caused the problems, and neither does the study suggest so, but they do not detract from the issue that they occurred with the biodiesel vehicles (Buchholz et al., 2005, p .4–5).

In addition, the EPA established the first alternative fuel volume mandate. The mandate set minimum volumetric requirements for alternative fuel to be blended with traditional petroleum based fuels, increasing over time to 7.5 billion gallons by 2012, and was later expanded by the EISA in 2007 (Frequently Asked Questions—EERE, 2009, p.2). This mandate is the primary driving factor in the rapid expansion of biofuel availability in the United States.

The EISA of 2007 was enacted with the purpose of improving vehicle economy and reducing dependence on petroleum. This act increased the mandatory alternative fuel volume to 36 billion gallons by 2022, established new categories of alternative fuels (setting volumetric requirements for each), and required the EPA to use a lifecycle analysis in the examination of alternative fuels (Alternative Fuels and Advanced Vehicles Data Center: Key Federal Legislation, n.d.). A lifecycle analysis requires consideration of the greenhouse gasses emitted throughout the production and use of a fuel, rather than simply the tailpipe emissions. The Energy Independence and Security Act of 2007 (EISA) amended the Renewable Fuels Standard (RFS), which established minimum volumes of ethanol to be used in the United States (How Ethanol is Made, n.d.). The current maximum blend of ethanol in regular gasoline in the USA is 10% (commonly referred to as E10), with variations up to that maximum chosen by individual fuel producers (Impact of Ethanol Blending, 2008). Manufacturers who sell cars in the US are legally required to cover under warranty blends of up to E10 (Impact of Ethanol Blending, 2008), and some manufacturers produce vehicles (commonly called flexible fuel vehicles) which are designed to run on E85 (85% ethanol 15% gasoline.)

The American Jobs Creation Act of 2004 emplaced additional tax incentives for biodiesel. These credits allow $1 per gallon of agri-biodiesel, and $.50 per gallon of waste-grease source biodiesel (Urbanchuk, 2010, pp.29–30). The Biorefinery Assistance Program, administered by the US Department of Agriculture's Office of Rural Development (created by 7 U.S.C. 8103), was established to promote biodiesel and other biofuels. This program "provides loan guarantees for the development, construction and retrofitting of commercial-scale biorefineries that produce advanced biofuels" (U.S. Department of Agriculture Rural Development Program, n.d.). Loan guarantees of up to $250 million, and grant funding of up to 50% of a project's costs are available through the Biorefinery Assistance Program.

Responsibility for oversight and application of biofuel programs is spread across a variety of federal agencies including the Environmental Protection Agency, the US Department of Agriculture, and the Department of Energy. The programs administered by the various agencies provide tax credits to producers, create a research and information center, set volume requirements that create a mandated market for biofuels and alternative fuel vehicles, and supervise a broad spectrum of grant and loan programs for the creation and development of facilities and technologies in the biofuels field.

CASE STUDY—EXAMINATION OF CONTROVERSIES

Opponents of biofuels cite several interrelated arguments against them including: initial carbon release from clearing land (carbon-debt,) rising food prices due to demand for source crops, questions regarding actual greenhouse gas (GHG) emissions measures, and the inability of biofuels to supply a significant portion of demand for fuel. According to economic theory, increased demand for source crops (such as soybeans for biodiesel and corn for ethanol) leads to an increase in the price of those goods. The increase in basic staple crop price corresponds to increases in staple good costs, and livestock feed prices. An increase in price can lead to more land clearing to raise more source crop. Some arguments have been raised against biodiesel on this factor alone, claiming that the clearing of land releases more CO_2 than is absorbed by the crops for many years. This argument is commonly referred to as "carbon debt."

Studies have found that the conversion of land to crop production can result in a release of carbon, which is not balanced by the use of biofuels for extended periods of time (Fargione et al., 2008). In clearing non-agricultural land, CO_2 is released through burning and microbial decomposition. Fargione et al. found that the repayment time of carbon debts, incurred by clearing land, varied from decades to centuries. The repayment time was found to be highly dependent on the type of land converted, mostly from marginal and abandoned cropland to tropical rainforest, and the type of biofuel source crop raised on the land.

For marginal and abandoned cropland converted to biomass ethanol production, the debt approached zero years, while peatland rainforests in Malaysia converted to palm biodiesel production incurred a carbon debt of 423 years (or longer). These carbon debts, however, are not incurred if existing productive or fallow farmland is converted to biodiesel source crop production. Most agencies have reported lifecycle analyses using existing productive farmland, yielding more positive impacts from biodiesel use since carbon debt is not incurred. In summary, so long as fallow or existing cropland is used for production, the carbon emissions reductions commonly cited as arguments in favor of biofuels are valid. If the land is converted however, the impact has generally not been reported accurately, and may often be negative.

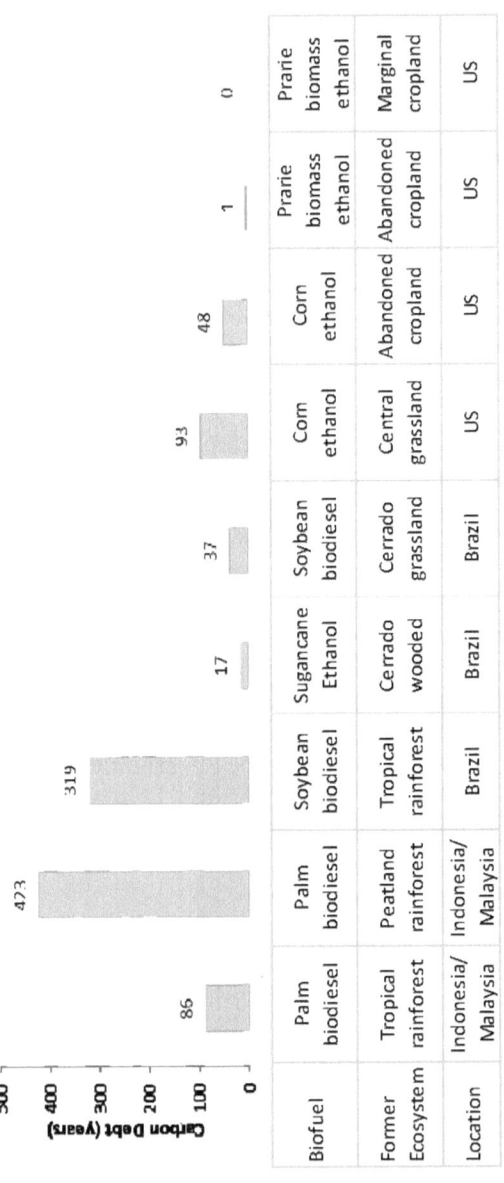

Source: Fargione et al. (2008).

Figure 8.1   Carbon-debt of biofuels.

As biofuel production expands, it is inevitable that land will either be converted from food production to fuel production, or that new land will have to be converted to production. As legislation mandated demand expands, and world demand for fuel of whatever type increases, it is clear that previously unproductive land will eventually be cleared incurring more carbon debt. This process is already underway in Malaysia and Indonesia, with rainforests being cleared for palm oil production (Young, 2011). Between 2005 and 2010 nearly one-third of Malaysia's rainforests were cleared, largely for palm oil production (used in various other consumer goods, but a chief source of biodiesel for the European Union). Indonesia and Malaysia account for 86% of palm oil produced in the world time (Fargione et al., 2008). Clearing of peatland rainforests for palm oil production incurs the highest carbon debts observed in Fargione's study of 423 years, but numerous other converted ecosystems also incur high debts.

Arguments for expanding production of biofuels through mandates, subsidies, and protectionist policy typically center on energy independence and reductions in carbon emissions. As the first argument is largely political in nature, we will not address it in here. The second argument is clearly flawed when the effects of carbon debt incurred through clearing land are taken into account. A positive net effect may be observed in some cases (or the debt quickly repaid) but even the average repayment time makes biofuel, as it is produced today, a very dirty "green" energy for many years.

Arguments can be made against biodiesel on a variety of different fields. Price is the major opponent when talking about the negatives to biodiesel. As there is an increased demand for corn and other biofuel crops, this causes the price of those crops, and consequently food, to increase, directly affecting both producer and consumer. According to economic theory, increased demand for source crops, such as soybeans for biodiesel and corn for ethanol, leads to an increase in the price of those goods. This price increase of basic staple crops increases food costs both of the staple good and in livestock feed prices; an increase in price can lead to an increase in clearing land to raise more of the source crop, then resulting in deforestation. This argument has been supported, but not fully quantified, by the International Monetary Fund (Mercer-Blackman, Samiei, & Cheng, 2007) and the World Bank (Zoellick, 2008), among other organizations (Alexander & Hurt, 2007). These organizations generally suggest that while farmers respond to higher demand by increasing production, in many cases, this increased production comes at a higher marginal cost. The social cost of subsidies to the various producers is not included in their analysis, which would suggest a higher real social cost than is incorporated in the price of the source crops. In many countries the increase in prices of various staple goods constitutes a significant portion of income, and has been shown to vary with the level of biofuel production (Ajanovic, 2010). The variance of food prices and biofuel production may also be explained (or influenced) by a number of other factors including agricultural subsidies, petroleum oil prices, international trade restrictions, and systematic market volatility (Ajanovic, 2010). Rising demand for corn (and other agricultural goods used in production of biofuels) has led to debates of "food versus fuel." Corn and soybeans are in demand both for human and livestock consumption, and some are concerned that the demand created by programs promoting biofuels will lead to higher food prices and unintended consequences throughout the agricultural and environmental sectors (Davies, 2007). Rapid expansion of ethanol use in fuel, estimated to reach 13.2 billion gallons by 2010, and the rising percentage of national production of corn used in the ethanol production industry (approximately 39%) likely have some inflationary effects on prices in the agricultural sector. The numerous factors influencing food prices and unclear data leave this particular argument impossible to evaluate at this time from anywhere beyond a theoretical standpoint (that increased demand will lead to increased prices).

A study commissioned by the International Energy Forum raises a number of questions, which further call into question the advisability of expanding biofuel programs (Mandil & Shihab-Eldin, 2010). This study suggests that the net effect of biofuels on GHG reduction may be marginal at best, and in some cases, unfavorably balanced, as the net effect depends highly on source crops used, geographic location, agricultural practices, and technology used to convert the source crop to fuel. Moreover, the study questions the economic viability of biofuels by posing the question: if the goal is to reduce GHG, is biofuel the best approach economically, or are there other, more cost effective approaches? Water consumption, reductions in biodiversity, and the GHG balance of fertilizer and pesticide consumption in biofuels are also significant impacts whose extent is not generally fully assessed.

In addition to the previous arguments, there are questions regarding the ability of biodiesel (and other biofuels) to satisfy a significant portion of market demand. According to the AFDC, enough of a variety of sources for production of biodiesel are available in the United States to produce up to 1.7 billion gallons annually, which would supply roughly 5% of the diesel used (Bozell, et al., 2004).

Reports by NREL suggest that biomass oils could supply up to 10 billion gallons (by 2030) with incentives and mandates, in a variety of forms. This figure includes, but is not wholly comprised of biodiesel; biomass oils can refer to a variety of forms of fuel derived from biomass, which includes ethanol. The question of the advisability and economic feasibility of achieving 10 billion gallons annual production was not fully addressed in this, or other studies, nor was the impact on consumers caused by the tax and transfer systems, compounded by the reductions in fuel efficiency. Furthermore, to expand production with the current source crops, new land would have to be cleared (see carbon debt discussion), and pressure on the supplies of those crops for food would have to increase sharply.

Finally, while CO2, SO2, and particulate matter emissions are generally reduced by biodiesel use, emissions of NOx are generally higher, a fact that clouds, to some extent, the balance of emissions benefits. Reductions in SO2 are unquestionably desirable, but the net effect of reduced CO2 with increased NOx leaves the matter in question. Another barrier to adoption is biodiesel's solvent properties. Although highly beneficial in increasing the lubricity of diesel engines, solvents tend to free deposits in engines converting to biodiesel use. These freed deposits can potentially clog fuel pumps and filters, which can be expensive (but by no means insurmountable) issues to resolve. Furthermore, the solvent properties of biodiesel tend to degrade rubber compounds commonly found in seals and hoses in engines, making conversion of existing diesel engines for use of blends over B5 problematic. This problem could be addressed in the manufacturing process and is by New Holland, which approved the use of B100, the purest biodiesel available. While conversion of all existing diesel engines for biodiesel use may be unfeasible, production of all new diesel vehicles to be compatible would not be a difficult task.

Fuel efficiency reductions also mitigate some of biodiesel's beneficial qualities. If the MPG of a vehicle is reduced by 10%, then more fuel will be used. The increased total volume used will generate an associated increase in total emissions, reducing the net benefits. This also comes at a higher cost as biodiesel carries a higher price than its equivalent in petrol diesel (both to the consumer at the pump, and the public in taxes to support subsidies.) This issue is equally applicable in ethanol use. Ethanol has been demonstrated to reduce tailpipe emissions of CO2 (as compared to gasoline), but clarity is lacking when the entire lifecycle and reductions in fuel economy are included. Like biodiesel, ethanol generates less energy when burned than traditional petroleum fuels; this results in lower miles per gallon.

Simultaneously, ethanol carries a higher price than gasoline. The combination of increased cost and reduced efficiency leave legitimate questions as to the impact of mandated increased usage on consumers. Additionally, subsidies to ethanol producers, coupled with protectionist tariffs on imported ethanol and the reductions in fuel economy, leave a great deal of doubt regarding the true costs of biofuel use.

These arguments leave a plethora of interrelated questions regarding both the true effect on national greenhouse gas emissions of biofuel use, and the cost effectiveness of achieving production and use levels that seem to have taken on a congressionally mandated, grail-like status among a variety of energy related government departments. If the goal is mass production of biofuels for the sake of producing biofuel, success is quite eminently possible; in fact, the US already leads the world in production capacity for both biodiesel and ethanol. This statement is of capacity, not actual production. The financial crisis of 2008 appears to have caused a sharp reduction in actual quantity produced from the previous trend. If the goal is to reduce greenhouse gas emissions, however, it is unknown if subsidization of, and a mandate for biofuels is a cost effective approach (or a carbon reduction effective approach). Studies have yet to be done quantifying the net costs and benefits, fully incorporating all the factors discussed above, in comparison with other approaches of achieving similar levels of GHG reduction.

ANALYSIS

Since there have not been any controversial cases regarding environmental institutions and biomass, instead we will focus on analyzing the underlying conflicts between government and biofuels. In the push for alternative sources of energy that can be found in country the US government, at all levels, has attempted to create an incentive system to create a biofuel market. These incentives, however, had many unintended consequences that followed the rush to create ethanol, biodiesel, and other biofuels.

POLICY ARENA

Issues dealing with biofuels take place all over the world. There are biofuel makers in many countries besides the United States, most notably Brazil. Even though biofuels have been in use for quite some type, their widespread use, in the United States, is relatively new. Thus, policies intending to cultivate their broad use are also new, causing unexpected consequences.

Policies hoping to either encourage citizens from producing biofuels vary from country to country, creating incongruities in the market. A feature not seen in any of the other alternative energy sources we explore, because the other alternatives are not easily traded. The components from biofuel can be grown anywhere, and often will be grown where it is cheapest to grow them. Because of this imbalance between the costs of labor between countries, ethanol production will move to areas where it is cheaper to produce and be sold to countries where the price is higher. In the process of growing the components for biofuels, land will be cleared, causing the release of carbon into the atmosphere, and the destruction of trees that help remove that carbon. Along with this problem, farmers move from growing food to growing biofuel components, when the price those ingredients fetch is higher than the price for foodstuffs.

KEY STAKEHOLDERS

The first and most important stakeholders are worldwide farmers. Whether farmers are cultivating a crop used in biofuels, or not, the demand for biofuels will affect their lives, hence they have a stake in biofuels. Those farming biofuel ingredients will see a higher demand for their product and will act accordingly. Those not farming biofuels components ingredients might either switch to farming biofuels components, or wait to see if there is a rise in demand for their foodstuffs as other farmers switch.

The stakeholders that set the guidelines this market functions are the politicians, specifically the politicians in America, and other countries where there is a high demand for this fuel. These men and women set the tax policies that decide whether it will be economical for biofuel ingredients to be imported or not. Additionally politicians, specifically Congress can decide to grant subsidies to farmers who are producing corn for ethanol, or beans for biofuels. These subsidies can then affect market supply and price, across the globe.

RULES IN USE

Federal tax credits, subsidies and mandates all affect the production and consumption of biofuels. One of programs covered earlier in this book is the PACE program. Known as the Property Assessed Clean Energy program allows local governments, after state authorization, to help individuals finance alternative energy. Based on their real estate tax, individuals are awarded funding to help with alternative energy projects.

Additionally, the federal government has offered the corn ethanol industry in the United States many advantages. In 2011 the federal government offered the corn ethanol industry $6 billion worth of subsidies. In the thirty years previous to that corn subsidies totaled almost $45 billion (Watson, 2012). On top of this, a tariff was placed on all Brazilian ethanol, to allow the US industry to operate without competition (Watson, 2012).

INFORMAL RULES

One of the reasons that corn ethanol has gained such a foothold in the United States is because of how well it was marketed. As soon as gas prices would begin to rise people would begin to talk about switching to ethanol, biodiesel, and other biofuels. Corn ethanol has been subsided to the point where the cost of pumping an ethanol-blended fuel into your car about equals that of pure gas. Gas stations advertise that they offer ethanol blend fuels, as one of the many ways of attracting customers.

The corn ethanol industry has been able to make people feel good about what they put into their car. Because people suddenly feel as though what they are putting into the car is not quite as bad as it could be, and so they continue to demand ethanol. We consider this to be an unspoken rule, because for many people debating the disadvantages to ethanol is unacceptable. Additionally, many people simply do not understand the far-reaching effects of what is sitting at the gas pump, waiting to be pushed into their tank. Advertising can create cultures and unspoken expectations; it has become an informal rule and agenda setter in modern society.

OUTCOMES

The impact of subsidies, tariffs, and mandates on countries around the United States has been great. Cheap, subsidized US corn floods local markets, desecrating local production. Industrial farmers are able to make large profits, while Americans end up paying more for biofuel than they would if it were imported from countries like Brazil. Additionally, the federal bankrolling of the corn industry distorts the market. Instead of looking for more viable, cheaper, and more efficient biofuels, the focus remains on producing corn ethanol. As mentioned above scientists could be working on other biofuels that would make a greater impact on lowering carbon emissions.

FUTURE

Biofuels have existed since the introduction of the internal combustion engine, but have not comprised a significant market share due to their higher production costs and lower efficiency when compared with traditional fuels such as gasoline. Numerous federally administered tax incentive, grant, and loan programs have effectively lowered the cost of biofuels at the pump (at a social cost through taxation and transfer.)

Although the benefits of biofuel use have been established regarding lowered tailpipe emissions, we see that biodiesel use significantly lowers CO_2, SO_2, and particulate emissions as compared to engines operating on petrol diesel, but results in higher NOx emissions. Ethanol use results in lower CO_2 tailpipe emissions than gasoline. Many vehicle models have demonstrated significant reductions in carbon emissions with "cradle to grave" analysis, the issue is somewhat clouded in regard to the carbon debt left by clearing land to initiate production of biofuel source crops and the carbon impacts of fertilization and pesticides. This carbon debt approaches zero for land already in active production of crops and conversion of fallow cropland. The carbon debt is extremely high however, when rainforest is converted to production of biofuel source crops. This is commonly the case in the production of palm oil for conversion to biofuel, which has accelerated in countries such as Malaysia and Indonesia. Most reported life-cycle analyses of carbon impacts assume conversion of existing productive farmland, which generally yields positive carbon impact from biofuel use. This method of analysis is not always indicative of the true carbon effects of the use of biofuels, and has led to overstatement of the benefits.

The carbon debt issue alone makes the effectiveness of biofuels in reduction of greenhouse gasses, extremely questionable. Even the best case scenario, assuming continued production of biodiesel and ethanol for the full repayment periods, leaves a significantly negative carbon impact in the interim. If alternatives to these fuels are found, and the land falls into disuse, the carbon debts would not be repaid. In this case, "green" energy and "green" outcomes seem truly at odds.

Developments in production of both biodiesel and ethanol may lead to a more economically viable alternative fuel in the future. Increased efficiency in the production, or refinement processes could be one avenue to achieve a price effective product. Alternatively, some issues may be avoided by producing fuel from different sources. Production of biodiesel, in particular, could become a more effective alternative fuel by utilizing different sources. At present, issues exist in regard to demand driven price effects on staple crops, and carbon debt, both of which could be avoided by finding a different source. That source may already have been found; biodiesel can be produced from particular kinds of algae, which are high in lipids. This source exhibits potential in several respects; the types of algae used can yield higher energy production per acre, can be used to remove impurities in waste water, would remove the increased demand for staple food crops, and can be processed in a manner very similar to current biodiesel transesterification (Biodiesel from Algae, n.d.). These factors make algae based biodiesel an excellent replacement for biodiesel sourced from food crops. The drawback, however, is the same as with ethanol and biodiesel, it is not price competitive with petroleum-based fuel at present.

At this point, with federal subsidies and mandates, biofuels have attained a degree of near competitive pricing at the pump, and a rapidly expanding productive capacity. The net balance of social cost of subsidies and benefits of biofuel use is uncertain, as is the true impact of use of biofuels on net carbon release and the cost effectiveness of the benefits of biofuel compared to other avenues that achieve similar outcomes (if any). As congressionally mandated, biofuels will see continued use, and expanding production. Advances in production processes, or development of other sources (such as algae) could change things, but as it stands, biofuel is a rather dirty "green" energy.

9 Oil Shale

Until quite recently, the Uintah Basin located in eastern Utah, has passed through history as little more than a blip on the map, occasionally warranting a sentence or two in a history book or pamphlet. And yet, a seemingly innocuous event, begun millions of years ago, created a chain of events that now seem determined to pluck the area out of obscurity and plunge it into the middle of one of the most hotly discussed political topics of this decade. The event in question was the creation of the prehistoric Uintah Lake. The lake formed during the Pliocene Epoch Tertiary period which lasted from 5 to 1.8 million years ago (San Diego Natural History Museum Fossil Mysteries, n.d)

The formation of this lake led to the slow deposit of sediment along the bottom of the lake, which began a slow geological process eventually culminating in the formation of large deposits of gilsonite, tar sands, oil, and oil shale under the basin through a process known as diagenesis (Uintah County Profile, n.d.). Formed from the remains of dead fish, plankton, and other organisms found in the water, organic aquatic sediment settled on the bottom of the lakebed. Once the lake disappeared and layers were piled upon this sediment, it was then subjected to a process of compaction and heat. Water from this sediment was pushed out through chemical reaction, compaction, and microbial action. Proteins and carbohydrates were then broken down and formed new structures such as kerogen and bitumen, the building blocks of oil, coal, and other fossil fuels.

Oil shale by definition is a form of sedimentary rock that contains a solid bituminous material (sometimes known as kerogen) that releases petroleum-like liquids when it is heated. A close cousin to oil shale is what has become known as tar sands, which is by definition a type of sandstone from which lighter fractions of crude oil have escaped, leaving residual asphalt to fill the pore space. Currently, oil shale and oil sands are considered by many to be a promising future source of petroleum, but they are also considered by an equal number of people to be dirty forms of energy with numerous negative environmental consequences. Due to both the controversy surrounding the resources as well as the economic realities of its use, to date it remains a largely untapped and unused source of petroleum and energy. The process that created oil shale is quite similar to the one that created regular oil, but the amount of heat and pressure applied were not as high. Typically, oil shale contains enough oil that it has the potential to burn without any processing or alteration (Oil Shale and Tar Sands Information Center, n.d.). This fact can also be seen in water supplies that are used in the extraction and processing of oil shale, with the water itself containing enough of the byproducts to actually burn.

The vast majority of oil shale deposits in the United States are located in the Green River sedimentary formation. This deposit spans across a large area found in eastern Utah, western Colorado, and southwestern Wyoming. About 72% of this 16,000 square mile area is federally owned and managed which provides the barrier of federal regulations to any attempt at the use of this resource. A significant portion of this shale has been set-aside for Navy use beginning with the passage of the Picket Act of 1910 (Andrews, 2011, p.8). This was done primarily in the hope that one day this shale could supply the Navy's oceanic fleets with its supply of necessary petroleum (Andrews, 2011, p.2). The Green River formation has been estimated to contain more than 8 trillion barrels of shale oil, of which about 1.8 trillion barrels appeared to be attractive to production. Much of the formation has been deemed either not recoverable or unusable, due to it either being too thin, too deep, or too low in given existing technology (Andrews, 2011, p.1).

Oil sands or tar sands are a combination of sand, clay, water, and bitumen, and is similar to oil shale. It, much like oil shale, can be mined and processed to extract the bitumen, which then can be further processed into usable oil. Tar sands are found in small quantities throughout the world, but the largest deposits are located in Alberta and Venezuela. The largest oil sand deposits in the United States are located in the eastern half of Utah, with the majority existing upon public lands. The oil sands in eastern Utah are estimated to contain somewhere between 12 and 19 billion barrels of oil.

Since the discovery of oil shale as a possible alternative to regular oil production, there have been numerous attempts to begin production and development, but there has yet to be a truly successful attempt at doing so. The early 1960s saw exploration and development of oil shale resources by several major oil companies begin. For example in 1961, the Union Oil Company began testing a retort at Parachute Creek, Colorado. This retort was shut down a mere eighteen months after it began operations due to the high costs involved in extraction and processing (Andrews, 2011, p.9). Following several unsuccessful attempts at oil shale production during the 1960', 1970's and 1980's, interest in the resource gradually faded. This all changed in 2005, when Congress conducted a number of hearings on oil shale to examine opportunities which would allow for environmentally friendly development of oil shale and oil sands to begin (Andrews, 2011, p.12). On January 17, 2006, the BLM announced its acceptance of eight proposals from six companies to begin development of oil shale related technologies, marking the first step towards widespread commercial use and development of oil shale in decades (Andrews, 2011, p.12).

When compared to regular petroleum and other fossil fuels, oil shale can be looked at as "teenage oil" (Vawter in Roberts, 2011). This is because shale comes in the form of kerogen, the first stage of organic matter's metamorphosis into petroleum (Andrews, 2011, p.1). This first stage of the transformation is called diagenesis. It occurs at a relatively low temperature, and occurs when an organic composition of nitrogen, oxygen and sulfur are released from the organic material trapped in the rock bed. The next stage—catagenesis—leads to the creation of hydrocarbons, and some natural gasoline. This stage requires a steep increase in both pressure and depth, when compared with the diagenesis state, to allow methane gas and graphite to form (Andrews, 2011, p.3). Oil shale, however, is not buried far enough below the earth's surface to allow this second stage to happen. Thus, it remains stuck in the diagenesis stage and exists as a lesser-developed form of petroleum.

Mining for natural gas from shale began in the 1820s in western Pennsylvania (Shale Gas Timeline, 2011). It wasn't until over a century later however, that the first natural gas boom occurred. As people began a general migration into cities the demand for cheap energy increased, as a result, natural gas began to play a vital role in the supplying of this energy. By the 1940s hydraulic fracturing (hydro fracking) came into existence as a method for the extraction of natural gas from shale. In this process water is pushed into the small cracks under the earth where natural gas sits, fractures the rock to release the gas, and then pushes it to the surface (Shale Gas Timeline, 2011).

Mapping techniques were significantly improved by the 1980s, allowing scientists to discover more deposits of natural gas, which were lying deeper under the earth's surface (Shale Gas Timeline, 2011). During this time scientists had also discovered another effective drilling technique, known as directional drilling. Also, massive hydraulic fracturing was patented and began to see more widespread use during this time period. By the 1990s the US government began to extend tax credits to those looking to explore and mine for shale gas (Shale Gas Timeline, 2011). By the beginning of the twenty-first century, a full-blown natural gas boom began again. Today, conservationists argue in favor of natural gas, when faced with a choice between it and other forms of fossil fuels such as oil or coal. All of these advancements made between the 1940s and today were largely as a result of to the support offered through the government, whether it occurred through government grants, research, or public/private partnerships (Trembath, 2011).

Currently, oil shale production is non-existent on any commercial scale largely due to the current high cost of the processing required to transform it from its original sedimentary form into usable oil. The potential for oil shale use has been a point of interest to oil companies for decades, with some production beginning in the late 1970's due to the energy crisis. This production ceased almost immediately following the end of the crisis, due again to the high cost of processing the oil shale into a commercially usable form. As a result of this halt in production, the technology used to process the shale has changed very little in the last thirty years due to the fact that oil shale has remained, simply a non-factor in the energy world. It has only been in recent years that the price of foreign oil has increased exponentially to the point where oil shale has nearly become an economically viable option. Perhaps as technology continues to improve, oil shale production will begin to become cheaper and thus become a more viable economic option as an energy source (Oil Shale and Tar Sands Information Center, n.d.).

The situation regarding oil sands is somewhat similar to that regarding oil shale in that neither source is currently utilized upon a commercial scale in the United States. Currently, the only nation that has a commercially developed oil sands industry is Canada. Canada produces roughly 40% of its oil directly from oil sand deposits located in northern Alberta. Roughly 20% of United States crude oil originates in Canada, which would indicate that a potentially sizable portion of the oil imported by the United States, comes from oil sands (Tar Sand Basics, n.d.). It is interesting to note that it takes roughly two tons of oil sands in order to produce one barrel of oil. Roughly 75% of the bitumen can be recovered and used from oil sands, the remnants are then discarded. Because there is currently a major commercial operation in Canada for the production of oil sands, it is therefore feasible that the oil sands found within the United States would be potentially useful in a similar manner. But one area of the Canadian operation which no doubt would translate into the United States is the fact that the use of oil sands exists as one of, if not the dirtiest forms of energy production in the world. This simple fact alone basically guarantees that there will be no expansion in this area of energy production within the United States until a method of development arises which increases the cleanliness of this resource significantly.

The current price to convert oil shale into standard oil is higher than sixty dollars per barrel, which emphasizes the main barrier to commercial expansion into this area. Despite this current high price of oil shale production, there is great potential for improvement, primarily through the advancement of technology that would reduce the price of production. One such area that hold potential for future advances is a process that is known as surface retorting. As the name suggests, surface retorting is simply the process of taking the oil shale directly from the source and feeding it into a kiln on the surface to extract the oil. The other advancement which holds high potential is a process which is known as "In Situ" retorting, a concept that is currently under development by Shell Oil. This process involves placing an electric heater down vertical holes that are drilled into a section of oil shale. This oil shale is then heated for an extended period of time (two to three years) until it reaches a temperature of roughly 650–700 degrees Fahrenheit. Once this temperature is reached, the oil is released from the shale and gathered in collection wells positioned near the heating zone. This plan, according to Shell, involves the use of ground-freezing technology that would be used to establish an underground barrier around the extraction zone. This "freeze wall" would prevent groundwater from entering the extraction area, all the while keeping the contents of the extraction zone from escaping and causing damage to such areas as groundwater supplies. The technology that Shell has developed is currently unproven on a commercial scale, but it is one that the U.S. Department of Energy regards as a very promising future advancement. The process is currently undergoing confirmation of the technical feasibility regarding the concept. Two areas in particular necessitate further testing: Controlling groundwater during production and preventing environmental problems upon the subsurface level, the primary focus being groundwater impacts (Oil Shale and Tar Sands Information Center, n.d.). Both of these areas of concern directly deal with the freeze wall theory, which as of yet is one of the areas that has not been proven as an effective method. If this method can be proven feasible there are several distinct advantages of the use of this method as opposed to more traditional mining methods. Advantages of this method include a significant drop in the amount of land disturbance required for extraction, access to deeper previously inaccessible shale reserves, and an overall higher quality in the oil produced. However, there are several disadvantages when compared with the more traditional methods of extraction such as the potential groundwater contamination previously mentioned as well as larger amounts of energy consumed to support the electric heaters.

Oil shale is mined in several distinct ways, of which two are very similar to the extraction method used to extract another fossil fuel, coal. The mining of oil shale is often done as a surface mining operation, which as the name suggests means digging down from the surface. This method is somewhat ineffective in the context of oil shale in that the largest deposits lie in areas where surface mining simply cannot access them. The second method, underground mining is, again exactly what the name would indicate. This method is similar to the method used to extract coal in which the mine is dug into the ground, and the shale is returned to the surface for processing. The third method that as of yet is unproven upon a commercial scale is the previously mentioned in situ method. Oil sands are mined in a manner nearly identical to those used for oil shale, but one difference can be seen in the use of the in situ method, which is already being used commercially for the extraction of oil sands.

The debate surrounding the use of oil shale is largely focused on the environmental concerns, which couple with the economic realities to present an interesting debate. The environmental concerns surrounding the extraction, processing, and refining of oil shell are similar to those present with other fossil fuels. The mining of this resource, particularly surface strip mines, leaves massive scars on the earth in a manner not unlike coal mining. These abandoned mines also tend to have large traces of potentially dangerous chemicals such as arsenic, located within the extracted and discarded shale.

Another area of concern for environmentalists is the fact that oil shale when converted to oil holds all of the same environmental impacts as traditional oil. This includes the greenhouse gases which are released when they are both burned, which many argue is one of the root causes of global warming. Another area in which many have expressed their concerns is the connection which oil shale production has with local water supplies. Many people theorize that any production of oil shale particularly in Utah will place strain upon the water supply, specifically that provided by the Colorado River. Not only will the development of oil shale have a negative impact upon the amount of water available, there also is the potential for the development of water quality issues. This fact is emphasized by the example given earlier of people being able to light their tap water on fire due to the presence of oil shale by products within the water, a sight that has become a YouTube sensation.

Many of the same problems have been seen in the extraction and production of oil sands, which if possible are even dirtier than oil shale. These facts have led many environmental groups, as well as many politicians to classify oil shale as a dirty form of energy that should not even be considered as a viable option. This does brings the economic side of the coin to prominence however, with the simple fact that the United States is already wallowing in a modern energy crisis. With the massive amounts of foreign oil being imported every day by the United States, it seems fairly obvious that some form of action must be taken to curb this exponential import trend. Proponents of the use of oil shale argue that its use would cut large portions from the amount of foreign oil which is imported daily and annually, which would in theory result in lowered energy costs. The recent events in the Middle East involving Iran's threat to block the Strait of Hormuz illustrates the dangers and uncertainty of continuing to rely heavily upon imported oil, primarily oil which sources from potentially unstable regions. If Iran goes through with this threat, it would cut off oil resources that the world desperately needs, not to mention potentially start another war in the region.

Another benefit of expanding into the area of oil shale production is the simple fact that it will create thousands if not tens of thousands of jobs in an economy. Opponents counter this argument by maintaining that the technology currently available is not suitable to accomplish production and usage of oil shale upon a commercial scale while doing so in an environmentally friendly manner. They argue that those in favor of oil shale usage are simply far too eager to begin mining, and are failing to consider the environmental realities of any such action.

Another area that has great potential for the expansion and improvement of shale extraction can be found in the area of what has come to be known as hydraulic fracking. This method, while effective is also highly controversial due to a lengthy list of environmental impacts in addition to its numerous health and safety issues. This process allows for the extraction of resources such as oil sands, shale, and coal formations that normally wouldn't be considered suitable candidates for extraction. The process begins after a steel pipe has been inserted into a well bore, which is followed by an injection of fracking fluids. This mixture of water and chemicals is pumped directly into the formation that leads to a buildup of pressure to the point where the rock cracks. One of the numerous negatives which was aforementioned is the issue which arises when looking at the water usage of this process. Due to the nature of the process a large amount of water is injected into the ground along with a small (in proportion) amount of chemicals. A portion of these fluids does eventually flow back up to the surface, however it is estimated that anywhere from 20–85% of fluid remain trapped in the ground. This can obviously prove harmful to the surrounding environment by contaminating the ground water supply and leaving large amounts of toxic chemicals trapped within the ground.

Predictably, many of the chemicals that are used in this process are highly toxic to both humans and wildlife with several of these having been directly linked to cancer. The process of rock fracking has several potential upsides most notably the improvement of the extraction process, but the environmental impacts and health and safety concerns certainly necessitate further research and technological advances to limit or eliminate these risks. The use of this method of extraction exists as one of the hottest environmental issues seen in today's political climate, but with such potential to expand the extraction of not only shale but also sands and coal deposits it merits closer scrutiny (Hydraulic Fracturing 101, n.d.).

There are many advantages to the use of oil shale, with the obvious leader in this category being the sheer amount of the resource located within the United States. It has been estimated that the amount of oil shale located in the Green River formation contains much more oil than the entire Saudi Arabian deposit, with it roughly estimated to contain between one and three trillion barrels of oil. Another advantage to the use of oil shale is the fact that it would be a domestically produced product that would supply the United States with not only energy, but large amounts of potential jobs as well. It is interesting to note how the expansion of oil shale production would affect the small communities in oil shale regions of Eastern Utah and Western Colorado. The rapid expansion of the oil shale mining operation in this area would dramatically increase the size of many communities, and no doubt would change the area both culturally and demographically forever.

REGULATIONS

There are a few federal regulations that correlate directly with the extraction and processing of oil shale. They all vary with purpose but impact the systematical approach to extracting and processing of the oil shale. The regulations are: the Clean Water Act, the Safe drinking Water Act, the Clean Air Act, and National Environmental Policy Act.

Clean Water Act: All waters in the United States are regulated through this act. It establishes the basic structure for regulating discharges of pollutants into the waters, and regulating quality standards for surface waters. Section 404 of the Clean Water Act requires a permit from the Army Corps of Engineers in order to release, fill, or dredge material in waterways; this includes wetlands into the definition of waterway. This affects the process of extracting oil shale directly due again to the reasons mentioned before regarding the affect which oil shale has upon groundwater and water resources such as the Colorado River. In particular the concept of in situ extraction is an area where the notion of preventing contamination of the groundwater in addition to other local water sources is most clearly seen.

SAFE DRINKING WATER ACT

The Safe Drinking Water Act requires the Environmental Protection Agency (EPA) to set standards for the quality of drinking water and oversee the states, localities, and water suppliers who implement them. The SDWA covers all drinking water and sources in the United States except for private wells and bottled water. The SDWA has been cited in litigation against hydraulic fracturing, or fracking, which is used to extract oil shale and other materials. The SDWA impacts oil shale extraction and processing in a number of ways, which all deal directly with the manner in which oil shale production depends upon water. Again, the in situ method of extraction is a cause for concern in that it involves the pumping of chemicals into the ground, of which potentially over three quarters of which could remain in the ground. Other concerns that this Act raises in the context of oil shale lie in the usage of water from external sources, in the case of eastern Utah, the Green and Colorado Rivers. With a large portion of these rivers being regulated to provide water to various states downstream, any dramatic increase in the amount of water necessary in Utah for such projects as oil shale processing, would conflict directly with this.

CLEAN AIR ACT

The Clean Air Act gives authority to the Environmental Protection Agency (EPA) to develop and enforce regulations to maintain air quality nationwide.

The way in which they achieve their goals is mostly through regulation fuel standards with their emissions. However in the case of oil shale, this could affect the hazards that are left behind during the extraction of it. The remnants left at abandoned mining sites provides an area of high concern not only under the context of this Act, but also in general. Other potential sources of conflict between oil shale interests and the CAA lie following its processing when it is used in a manner similar to that of normal oil which is burned in various applications.

NATIONAL ENVIRONMENTAL POLICY ACT

With its basic policy being the assurance that all branches are cooperating in environmental consideration (US EPA, n.d.), NEPA qualifies which projects are environmentally safe through Environmental Impact Statements (EIS) (National Environmental Policy Act (NEPA), n.d.). It requires agencies to disclose these impacts to interested parties and the general public, which then in turn could hamper any progress in any direction due to differencing parties (Arthur, 2008). Obviously this affects the use of oil shale resources due to necessity of any such action undergoing a NEPA analysis prior to any development occurring. The interesting fact surrounding any NEPA analysis and EIS surrounding oil shale development is the fact that oil shale has a terrible reputation for being a dirty source of energy. Thus, there may exist potential for conflict between regulation and mining interests in this regard.

CASE STUDY—UINTAH BASIN, ORION V. SALAZAR

In the Uintah Basin, the temperature and pressure eventually dropped below a critical point, and the process of diagenisis stopped; had this process continued, crude oil, coal, or natural gas would have formed beneath the basin. As it was, the sedimentary rock oil shale was formed instead. Though the oil shale was not really shale, and it did not contain oil, the kerogen found in the shale was a precursor to petroleum and could yield oil or natural gas if processed and distilled (About Oil Shale, n.d.).

While this potential "black gold" was forming underground, human life flourished above ground in the Uintah basin, oblivious to the slow, methodical geological processes occurring below the surface. After the last Ice Age, the area was inhabited by the Paleoindians, the Archaic, and later the Fremont Indian Tribes. Due to the climate during this time, which was colder and wetter than it is in the present-day, the Paleoindians flourished in the region from around 11000 to 6500 B.C.E. (Utah's prehistory in a nutshell, n.d.). The Archaic Indians succeeded the Paleoindians, and are noted for their "evocative rock art" (Utah's prehistory in a nutshell, n.d.). Evidence of the Fremont

Indians, who inhabited the area from around 500 B.C.E until around 1250–1500 A.D, includes their unique moccasins, clay figurines, basket weaving, and gray pottery, items which can still found in scattered ruins within the Uintah Basin; petroglyphs and pictograms also attest to the legacy of these prehistoric tribes (Dinosaur National Monument—Fremont, n.d.). These groups were followed by the Ute Indians, which consisted of twelve loosely affiliated bands. The Uintah Basin, Uintah Mountains, and Uintah County, as well as the state of Utah itself, all derive their name from the Ute Indians. In contrast to their predecessors, who had engaged in semi-permanent farming lifestyles, the Utes practiced a "flexible subsistence system"; because of their remote location, the Utes also had little contact with European explorers until the Dominguez-Escalante expedition of 1776 (Ute Indians, n.d.).

The modern, recorded history of the Uintah basin and the surrounding region begins, as many such stories do, with a journey. Father Francisco Atanasio Dominguez and Father Silvestre Velez de Escalante set out from Santa Fe on July 29, 1776, in hopes of finding a safe travel route from Santa Fe, New Mexico to Monterrey, California. They journeyed due north, avoiding the Chirumas cannibalistic tribes directly to the northwest, preferring the somewhat friendly Ute territory to the north. Their journey, which lasted over five months and covered 1700 miles, took them through previously uncharted territory in modern day Colorado, Utah, Nevada, and Arizona. Fearing an early winter, they turned back from Nevada to New Mexico rather than risk frostbite, cold, and starvation. While the expedition was unsuccessful at finding a route from Santa Fe to Monterrey, California, the priests kept detailed and thorough notes of the geography, landscape, and cultures they saw as they passed through the region. These notes and maps, illustrating the particulars of the region, detail the expedition of the first white men to the area (The Dominguez-Escalante Expedition, n.d.). Of particular interest is the note they made of the Uintah Basin area, dated September 13, 1776:

> There is here a fine plain abounding in pasturage and fertile, arable land, provide it were irrigated, which might be, perhaps, a little more than a league in width, and some four or five leagues in length, entering in between two mountains; the space taking the form of a corral, and the mountains coming so close together that one can hardly distinguish the opening through which the river flows. The river can be crossed only at the one fording place, which our guide assured us was in this neighborhood, to the west of the mountain that stood farthest to the north, close to a range of hills composed of loose earth of a leaden color, and, in places, of a yellowish tinge. The bottom is full of small stones, and the river so deep that the mules could not cross it except by swimming. (Harris, W.R., et al., 1909, p. 165)

European settlements in present day Uintah County gradually expanded following the expedition, as trappers began to discover such natural resources as beaver and other wildlife within the Basin. A small outpost built by Antione Robidoux, a French trapper, operated from 1831/32–1844; however, hostilities with the local Ute Tribes forced him to abandon the post. Problems between the indigenous population and Euro-Americans only worsened after the Mormon pioneers settled in present-day Utah near the Great Salt Lake in the Salt Lake Valley beginning in the summer of 1847. The Walker War (1853–54) was a direct result of increased white settlement and infringement upon traditional Ute lands.

In 1861, President Lincoln set apart two million acres for the Ute bands called the Uintah Valley Reservation; The same year, Brigham Young, leader of the Mormon settlers who had come to the area to escape persecution, sent a group of explorers to investigate the potential for possible Mormon settlement of the Uintah Basin area. After receiving the report that "all that section of country lying between the Wasatch Mountains and the eastern boundary of the territory, and south of Green River country, was one vast contiguity of waste and measurably valueless", Brigham Young decided against settling the region (Uintah County, n.d.). This changed when gilsonite was discovered in 1888, and miners quickly pushed for the Federal government to "withdraw" 7,000 acres from the Uintah Reservation so they could mine the gilsonite. The government quickly complied and the area removed from the reservation, called "The Strip", quickly degenerated into one of chaos and lawlessness similar to that seen in many other mining towns across the west (Uintah County, n.d.). Oil shale, used by early European settlers in the region as a substitute for oil, was also discovered in the basin during the same time period. The oil shale was heated until the rock began to secrete a dark, blackish-brown sticky substance; this substance was then used to lubricate machinery. Although development of the shale was attempted, it was quickly abandoned because it was not economically viable to produce the oil en masse (Roberts, J., 2011).

In order to make oil or natural gas from oil shale, the compacting and heating process begun during diagenesis must be artificially completed. Since humans do not have several thousand years in which to wait while natural processes would render the kerogen oil into usable form, the task is to find a process that will speed up this transformation. This goal is accomplished by heating the rock up to extract the oil and gas trapped within the shale; and though this process does produce the desired result, the technology and equipment necessary for such an operation creates massive overhead costs which has severely limited development (About Oil Shale, n.d.). The costs of extracting this shale oil and transforming it into usable petroleum translated into much higher costs for the finished product when compared to conventional petroleum, which does not require the processing and distillation that is seen with petroleum derived from oil shale. During the oil crisis in the late 1970s, early 1980s, and the subsequent skyrocketing prices of oil, production of oil shale was considered a viable option; however, the drop in worldwide oil prices in the mid 1980s cut that dream short. As concerns over geopolitical unrest in areas of present oil production increase such as the current situation with Iran, efforts to reduce foreign dependence on energy have sparked renewed domestic interest in oil shale development. In addition, increasing oil prices in the last few years have made the prospect of oil extraction from oil shale alluring in the hopes of cutting the high prices of traditional petroleum.

Enter the significance of the Green River Oil Shale Formation, found in Colorado, Wyoming, and the Uintah Basin in Utah. With between 1.2 to 1.8 trillion barrels of oil, it is easily the largest oil shale formation in the world. Of that amount somewhere between 750 and 800 billion barrels of oil can be recovered (About Oil Shale, n.d.; Oil Shale, n.d.). To put this amount in perspective, consider, for a moment, the proven oil reserves of Saudi Arabia; at 266.75 billion barrels, the potential oil in Saudi Arabia is less than one-third the size of the recoverable oil from oil shale in the Green River Formation (Country analysis brief: Saudi Arabia, 2010). At current levels of United States demand for petroleum, the estimated 800 billion recoverable barrels in the Green River formation would last more than 400 years (Bartis et al., 2005). With this being the case, there exists a possibility that oil shale might be the solution that has long been proposed by the peak oil dilemma. Extraction of large portions of this oil may prove difficult, if not impossible due to a variety of limiting factors. In addition to high overhead costs and technological obstacles another major roadblock lies in the fact that:

> More than 70% of the total oil shale acreage in the Green River Formation, including the richest and thickest oil shale deposits, is under federally owned and managed lands. Thus, the federal government directly controls access to the most commercially attractive portions of the oil shale resource base. (About Oil Shale, n.d.)

FEDERAL LAND MANAGEMENT

The United States has a long history of contradictory land management policies, broken down into four general stages of federal land policy: acquisition, disposition, retention, and management. In the acquisition phase, the goal was to acquire the as much land as possible on the continent through purchases, invasion, and settlement. This was done through enlargements such as the acreage gained in the Louisiana Purchase in 1803 which added roughly 523.4 million acres (Anderson & Martin, n.d., p 2), the annexation of Texas in 1845, and land acquired following the end of the Mexican American War in 1848. By the year 1853, the federal government owned approximately 613 million acres. This was further expanded by later conquests such as the Spanish American War in 1898 which added such areas as Puerto Rico, and other gains such as the purchase of Alaska from Russia in 1868 as well as the Gadsden Purchase of 1854 furthered this expansion. Once the federal government had acquired this public land, the goal of the government was to give away the land in order to populate it with "American" citizens during the disposition phase, protecting the newly acquired land from encroachment by other countries and empires.

Acts such as the Land Ordinance of 1785 made it clear that the federal government wanted to transfer federal land to private hands as soon as possible for to pay off the Revolutionary War debt, to secure the nation against invaders and to legitimize the United States claim to the land through settlement. The Homestead Act of 1862 illustrated a move towards the growing American notion of manifest destiny, essentially the notion of expansion across the continent. This Act allowed for the further divestment of lands by dividing land into 160-acre plots and making these plots available to settlers who cultivated and claimed ownership of the land. Disposition also occurred through the transfer of land from the federal government to the states as well as to private citizens. During the course of the nineteenth century the federal government sold approximately 871 million acres of federal land held in common to private citizens, a fact which emphasizes the point of how the federal government attempted to distribute the land (Anderson & Martin, n.d., p. 1).

While it continued giving this land away, federal officials began to question whether disposition of the land was be the best or most effective policy. A series of statutes were then passed with the goal to prevent the transfer of land, beginning what is known as the retention phase, moves that were backed enthusiastically by conservationists. This did not mean, however, that all economic activity and resource extraction on federally owned land ceased. Earlier American concepts of property held that property meant owning the physical plot of land with all rights to the land held exclusively by the landowner. The concept of property evolved during this time to mean that land could be divided into various different rights, such as: logging, mining, and foraging rights that could be given to citizens and corporations, while still allowing the federal government to maintain overall control of the property.

Public land reservation began with the 1872 creation of Yellowstone and continued with the General Revision Act in 1891. This act halted all public offering of land, slowed the transfer of title in the Homestead Act, and did not allow owners of more than 160 acres to receive homestead status; it also allowed the President to set aside public land covered with timber for protection. As more rules and regulations sprang up surrounding the use of public land, the transition of the federal government from realtor to manager of federal lands continued.

The federal government movement toward management and away from disposition really began during the environmental movement of the 1960s, with the enactment of legislation such as the Federal Land Policy and Management Act (FLPMA) of 1976, National Environmental Policy Act

(NEPA) of 1969, and others. These acts show a shift in the federal government's role in public land management from passive to active participant in public land management. The move from passive to active participant in the management of the land can also be seen in the creation and expansion of federal agencies commissioned to manage federal land as well as precedent established by court proceedings.

ORION V. SALAZAR-CASE STUDY

One noteworthy court proceeding demonstrating the active participation of the federal government in land management policy is *Orion v. Salazar,* the last in a series of cases involving 156 oil shale claims found on federal land in Uintah County, Utah. As a result of this case, precedent has been established that holds private leasers accountable not only for present as well as future regulations, but also in some instances holds them retroactively accountable to such regulations. But before going into the particulars of the case, a little history on established federal regulations is in order to better understand the background behind the lawsuit.

The Mining Law of 1872 allowed for extraction and mining of minerals on federal lands so long as $100 worth of assessment work was completed annually. Until the issuance of a patent, this annual assessment was a mandatory requirement of the owner of the claim. If a claimant failed to do the assessment, their claim became void and could be taken by a competing claimant; the "resumption provision", however, allowed for claimants to keep land rights if they resumed assessment work before another claimant staked a competing claim (Mineral Lands and Mining of 1995, Mineral Lands and Regualtions in General, 2011).

When the Mineral Leasing Act was passed in 1920, all previous regulations under the Mining Law were void, except for claims that qualified under the saving clause. This clause stipulated that claims made under the Mining Law were valid if all claims were "maintained in compliance with the laws under which initiated." (Mineral Lands and Mining of 1995, Leases and Prospecting Permits, 2011). This meant that as long as claims made before the enactment of the Mineral Leasing Act complied with regulations from previous acts i.e. the Mining Law of 1872, the claims would remain valid. One of the provisions of the Mineral Leasing Act is that claimants comply with federal, state, and local requirements in order to maintain their mining rights (30 U.S.C.S § 26, 28). One requirement of Utah law is that affidavits be filed showing that annual assessment work has occurred. These affidavits are considered "prima facie" evidence and must be filed within thirty days of the assessment in order to maintain the claim (Mines and Mining: Utah Code Ann. § 40–1-6(2–3)). Failure to file an affidavit does not necessarily result in the loss of a claim if other evidence exists which indicates that the assessment work was performed (Mines and Mining: Utah Code Ann. § 40–1-6(3)).

In 1988, owner Frederick H. Larson filed a patent application for 156 oil shale claims found between 1917 and 1919. Because of the large number of claims and area encompassed by the claims, the application was divided into nine separate applications at the request of the Utah BLM office. A lawsuit was filed in 1992 to compel the Secretary of the Interior to finish processing the applications. The court found in favor of Larson and ordered the BLM to continue processing the applications. Transfer of the claims from Larson to Crippled Horse Investments Limited Partnership occurred in 1994. In 1996, Crippled Horse Limited Investments filed another suit to compel the Interior to complete the patent applications, citing unreasonable delay on the part of the Utah BLM. The suit was later stayed on the condition that the BLM submit a progress report every sixty days.

The BLM voided several of the 156 claims in 1997 because of a persistent failure to do $100 of annual assessment work; although length of time varied, claims did not receive assessment work during the time period 1920–1970. This decision was delayed at the request of Orion, which asked that a decision be stayed until research had been completed on all claims. The BLM continued research on the claims and found that affidavits were not filed for fifteen to thirty-five years from 1932 to 1970. In 1999, the BLM declared all 156 oil claims void because of failure to complete assessment work, as evidenced by the lack of affidavits submitted for this time period. The decision was appealed to the Interior Board of Land Appeals (IBLA) in 1999, which affirmed the decision of the Utah BLM office. A complaint was filed in 2004 challenging the decision of the IBLA in *Orion v. Norton*. In 2006, the court decided in favor of Orion, arguing that

> . . .subsequent regulatory changes represent changes to the law plaintiffs (and their predecessors-in-interest) operated under previous to 1970. While it is certainly within the scope of the agency's informed discretion to update its regulations based on evolving interpretations of case law and statutory language, it is arbitrary and capricious to apply these new policy interpretations and regulations to past conduct because it forecloses the plaintiff's opportunity to avoid adverse consequences by operating in conformity with the agency's position."
>
> —Royce C. Lamberth (*Orion v. Norton* [2006])

For this reason, the BLM was ordered by the court to continue processing the 156 oil shale claims. In 2007, Orion brought another suit, this time seeking for an order to make the agencies process the patent applications or explain why the patents had not been processed within sixty days. Orion also requested that the court retain jurisdiction if the BLM failed to act within the sixty-day deadline. While the first request was denied, the court did agree to maintain jurisdiction over the matter.

The decision made in *Orion v Norton* was appealed to the Columbia District United States Court of Appeals in *Orion v. Salazar.* At issue was whether or not nullification of oil shale claims was within the jurisdiction of the BLM given that the annual $100 assessment work required by the Mineral Law of 1872 was not performed from 1920–1970, after which time the annual assessment work resumed. While Orion did not dispute the lapse of work, it argued that interpretation of federal law and precedence only required the resumption of work (as implied under the "resumption clause" in the Mining Law of 1872), and not the continuation of work on the claims, to maintain rights to the disputed claims. The Mining Law of 1872 was applicable because the claims in question qualified under the saving clause found in § 193 of the Mineral Leasing Act, which states that claims made under the Mining Law were valid if all claims were "maintained in compliance with the laws under which initiated" (30 U.S.C.S § 193). While more recent regulations and policies called for more stringent compliance with assessment work, such as those established by the Supreme Court case *Hickel v Oil Shale Corporation*, Orion argued that at the time these standards did not exist and the claims should not retroactively be held to a higher standard. In contrast, the BLM argued that "a lapse in assessment work causes the interest of the claimant(s) in the minerals subject to the mining laws to revert back to the public domain" (43 CFR § 3851.3(b)). Because there was a lapse in assessment work, the BLM argued that it acted correctly in voiding Orion Reserves Limited patents to the 156 oil shale claims.

The Columbia District United States Court of Appeals used the Hickel test established in *Hickel v Oil Shale Corporation* to resolve the issue. In *Hickel v Oil Shale Corporation,* the Hickel test established that "showing of intent is irrelevant when a mineral claimant has substantially failed to perform assessment work required by the General Mining Law of 1872" (*Hickel v. Oil Shale Corporation*, 1970). Using this reasoning, the Columbia District United States Court of Appeals reversed the decision made in *Orion v Norton*, arguing that the "saving clause" found in the Mineral Leasing Act did not apply in the situation because the claimant had failed the Hickel test by "substantially" failing to perform assessment work on the oil shale claims. In reversing the *Orion v Norton* decision, the Court of Appeals declared the BLM's decision to nullify the 156 oil shale claims valid.

ANALYSIS

Similar to the BrightSource narrative, here is another example of environmental laws being retroactively applied to leases. In this case leases granting access to oil shale on federal lands were held to requirements asking for a line in the General Mining Law of 1872 requiring that assessments be completed annually. Since these mining claims had not been actively used for several years, assessments were not completed. Thus the mining leases allowing access to potentially valuable oil shale were returned to the federal government.

POLICY ARENA

As mentioned in the Regulations Chapter (Chapter 2), federal mining acts were an attempt by the federal government to ensure their authority over the land they had acquired. The wave of miners heading across the West and mining on federal land were so sudden and unexpected that these acts were passed to prevent further encroachments and to allow the collection of at least some revenue off lands that were being mined. To ensure that these leases weren't bought and held eternally the federal government applied the stipulation that there must be an annual assessment.

At the time when these laws were passed, those writing the mining acts could not have anticipated the amount of oil shale and other minerals present, and future price of those minerals. Understand this can also help explain why the court order may seem inappropriate, taking away someone's lease rights for failing to complete and small (only $100) assessment. As such, whether by accident or not, a clause was built into the mining acts that would allow the federal government the opportunity to regain the ownership of those minerals. When oil shale becomes technologically and economically feasible it will mean the US government owns a large share of these oil sands.

KEY STAKEHOLDERS

The Bureau of Land Management, the agencies with authority over these leases would benefit most from regaining the leases. Either they could hold on to these leases until demand for oil shale increases, or they could resell the leases now for another profits, or to recover monies paid in legal fees. Additionally if these leases were in areas the BLM would like to see protected, they would first need to have ownership of the claims again, which they now have.

The other stakeholder in this case is the private individual, Frederick H. Larsen. Larsen, and later Crippled Horse Investments, would like to ensure they have access to their claims without unnecessary assessments, thus Larsen's decision to patent the claims. No matter the reason behind the change in ownership from Larsen to Crippled Horse Investments, both would benefit from having full access to their investment. Larsen must have thought that the leases would be able to bring him a good return on his investment, otherwise he would not have gone to all the trouble of taking the BLM to court to speed up his patents. It was only after ownership changed hands and the BLM was forced by the court again to pay special attention to these leases that they noticed the discrepancies.

RULES OF USE

The two laws at the heart of this dispute are the Mining Law of 1872 and the Mining Lease Act of 1920. As previously stated the purpose of these

Acts was to ensure that the federal government could grant limited access to its land, and maintain some level of control over that access. Although it was possible for any person to stake a claim on these leases once the previous leaser had failed to follow the guidelines established in the Mining Law of 1972, the government could easily step in, assess these lands and become the new claim owner. Generally this would be easier for the government than individuals because they readily have access to these lease claims.

Even the passage of the Mining Lease Act in 1920 did not override these claiming procedures, under the saving clause. Thus, the owners of claims would only have to prove they had done an assessment every year since they had owned the claim. The court's interpretation, however, did not agree with this logical argument.

INFORMAL RULES

The Columbia District US Court of Appeals made an unprecedented decision, one that was not written, or known beforehand. The court held that even if the Crippled Horse Investments or Larsen had intended to do these assessments, the fact that the assessments were not completed meant intentions were irrelevant. Thus, the claims no longer belonged to the Crippled Horse Investments Limited Partnership. Even though the rule the court used to make this decision, the Hickel test, was known beforehand, it was not known how that the court would use this rule, or interpret the rule in the fashion they did.

OUTCOMES

In this decision it was in the Bureau of Land Managements' best interest if the claims were returned to them. Otherwise they would have been forced to process the claims, with their associated patents, and they would have lost the rights to reclaim ownership under the assessment clause. The BLM was able to receive and unprecedented decision, one that will undoubtedly benefit them in the future as well.

The Crippled Horse Investments Limited Partnership would much rather have had the Columbia District US Court of Appeals stay the decision made by lower court. Instead the company lost the leases it had acquired, on top of the legal fees it had paid to get to this decision. Ultimately this decision furthered federal attempts to control their land and the minerals found in that land.

FUTURE

There is no doubt that fossil fuels play a major part in the United States today, with massive amounts of the infrastructure and transportation nearly exclusively dependent upon them. Currently it would appear that oil shale is an option that demands closer examination, but with better technology and methods necessary before it becomes a viable option. Oil shale can and will only play a major role in solving the United States' energy woes if technological advances make it easier and cheaper to process the shale. The only other instance in which it could become economically feasible is if the price of foreign oil continues to rise to the point where it becomes cheaper to convert oil shale into oil than to import the oil from a foreign nation. The other area surrounding the use of oil shale that necessitates further study is in the area of its direct environmental impact. The stories of water that burns raises questions as to the true impact which the production and use of oil shale will have upon the surrounding areas. There is no doubt that the economic concerns of oil and international politics insist that something must be done, but that cannot occur at the expense of the environment. As has been proven over the course of the last century, the extraction of fossil fuels can be done in a manner, which, though not good for the environment, still maintains a cordial balance between the two interests. Thus it can be concluded that oil shale, while having massive potential, is still only in the infant stages of its development as a viable resource.

The future of oil shale development depends upon two main factors, the first being the price of crude oil, and the second being the costs associated with the retorting and refining process. The past fifty years demonstrates a cycle, which illustrates the relationship between crude and interest in shale production. Each time oil prices soar, so does interest in expanding and continuing oil shale extraction and refinement as many seek a way to subvert the traditional import system. For example, the energy crisis of the late 1970's and early 1980's was marked by numerous oil company attempts to make oil shale economically viable. But by 1982, as prices began to decline so did interest in the oil shale industry, with nearly all oil company projects halting (Andrews, 2011, p. 13). Therefore, it can be inferred that this trend will continue into the future. If gas prices continue to remain high, so will interest in the oil shale industry as many search for a viable alternative the current import system.

10 Conclusion
An Analysis Using the IAD Framework

All of our case studies demonstrate the unintentional consequences that a myriad of environmental legislation, litigation, and regulation have created. These repercussions range from the global effects of biofuel subsidies distorting markets and moving foodstuffs from mouths into vehicles, to local interests groups' ability to bankrupt companies by way of legal fees, as was the case with the Calpine Corporation. Illustrated in these case studies is the continual use environmental preservation legislation by local interests to trump national or even global interests. While internationally the desire may be to switch to using alternative fuels, local communities do not want to bear the cost of siting and then maintaining these projects.

POLICY ARENA

Common among these studies are the conflicts that arise over common pool resources. Local interest groups again and again, have no desire to place these alternative energy plants in their community. Instead they free ride off energy siting by other communities, assuming that those communities are currently siting or will site alternative energy production. This however is too large an assumption. If it is the global interest to site these plants, but it is in no community's interest to have them placed there, alternative energy companies will be left to fight their way into a location. Even the federal government, which typically attempts to provide public goods, is forced to grapple with local interest groups when they try to place these sites on federal land.

These problems are further exacerbated by federalism. States such as California and Massachusetts have environmental laws that are stricter than their federal counterpart. Energy companies located in these states have to follow state regulations in addition to those federally regulations, to avoid costly litigation. This conflict can become even more inflamed when city councils layer ancillary regulations, some of which may be meticulous enough to prevent energy siting altogether.

KEY STAKEHOLDERS

One common theme among these narratives is that self-interested provincialism leads groups or citizens to oppose alternative energy plant sitings. Citizens do not want these plants built for fear of the imposition it will place on them. Not only would they be forced to incur the cost of siting and the possible loss of viewsheds, but also increased energy cost since these alternative energy sources often cost more to use, a cost that is passed onto consumers.

Because the actions of a single citizen are often not enough to prevent these projects from being sited, these individuals form coalitions. These associations then work to supply the desired outcome in exchange for the members' time and financial support. The members of these groups dictate the desired outcomes, and the demographics of their membership dictate the steps taken towards achieving those outcomes. The Alliance to Protect Nantucket Sound petitions local environmental regulators, state lawmakers, and national Congressional representatives. Because of the significant funding they receive, they are a powerful group. Smaller groups such as those writing the Mojave Desert Blog do not have the same means as the Alliance to Protect Nantucket Sound. Hence, their tactics rely more on making grand statements, and building grassroots support. Whether is it the Glen Canyon Dam Institute, or Pit River Tribe, these groups act because they believe the possible benefits of action outweigh the expected costs.

National environmental conservation groups will also chose to resist these projects when they see the benefits outweighing the costs. Cape Wind demonstrates national environmental groups backing an alternative energy plant, while the Glen Canyon Dam is strongly challenged by many national groups. The difference between these two is the difference between the perceived benefits and costs. With Cape Wind, interest groups see a much larger pay off for backing this plan, and little risk of losing support by backing it. The decommissioning movement, however, has grown over the years. In order to win support, and therefore donations, national groups have more of an incentive challenge these projects.

In the political arena, representatives will act in their self-interest. Often this means saying and doing what is necessary to be and stay elected. In Massachusetts, Deval Patrick was able to turn around Cape Wind's project after he was elected. His desire to push through this project was made known during his campaign, and didn't prevent his election. In Utah, there is a strong push for the state to control the land contained within its borders. Bills are constantly introduced in the state legislature demanding a redress for federal actions concerning public lands.

Utah's constant circulation of bills demanding action from the federal government isn't often met with results. Politicians have the incentive to make grand policy objectives while running for office. When it comes time to push all their objectives through, however, often it is difficult to follow up on those campaign promises. For many politicians, especially those on a state or local level, their power is limited. Even if local officials were voted into office because of their alternative energy platform, energy companies would still have to navigate the federal and state regulations. Politicians can appeal to other politicians and appoint more persuadable regulators, as was Deval Patrick's appointment of Ian Bowles. They cannot, however revamp the entire system individually, or immediately.

Regulators are incentivized to act in a way that allows them to stay in their job. For newly appointed regulators this means appeasing the politician who nominated you. Regulators who are seeing the guard change as a new politician enters office will want to prove their flexibility and willingness to promote the new administration's goals. When Ian Bowles was appointed, suddenly Cape Wind was able to get necessary state permits. In California, however, where state environmental laws are tougher, citizens elect politicians who maintain strict environmental standards, and the regulators must also remain tough. The California Energy Commissions' increased scrutiny between BrightSource Energy's two development projects demonstrates the change in public opinion and the response of the regulators to the resulting political changes.

One illustration of the conflict between federal, state, and local interest groups is Ivanpah Solar Plant. Julie Cart (2012) in a Los Angeles Times article outlined the deal negotiated among federal regulators, national environmental groups, and BrightSource Energy developers. National leaders cited the need to reduce carbon emissions; the possible destruction of local habitats and native plants was less harmful than the possibility of continued future carbon emissions (Cart, 2012). Local environmental group members expressed dismay over some of the permissions that Ivanpah was granted.

State leaders touted the jobs created, and the alternative energy that would now be available to its citizens. BrightSource Energy is able to site their plant, they have the option of siting future plants, and they have access to millions in subsidies and tax breaks.

RULES IN USE

Many of the rules that govern alternative energy siting are conflicting with political goals. Congressional legislation extending tax cuts and subsidies for solar development fails to take into account the Endangered Species Act, or the years it takes to go through the NEPA process. Again we emphasize that political goals such as the Department of Energy's goal of using 20% wind energy by the year 2030 overlook the necessary steps required before that energy can even be transmitted.

The network of environmental laws and intervening litigation has created a complex, dynamic system that is often used as a tool by interest groups, both local and national, to serve their interests. Those responsible for the Mojave Desert Blog would also like to be responsible for preventing the siting of the BrightSource SEGS. It is in their best interest to keep their viewsheds intact, protect the desert tortoise, and most important, prevent the BrightSource Energy plant for fear of it opening a 'floodgate' of solar plant proposals.

Politicians are incentivized to continue to pass environmental legislation as long as it is what they think their constituents want. The recent controversy over the implementation of stricter air pollution regulations by the EPA under the CAA signals the public might be trending away from increased environmental regulation, especially when it appears to harm job growth. Obama's support of the Ivanpah solar plant seems to indicate a national desire for more solar energy. Politicians have little incentive to do more than talk big about alternative energy as long as their constituents continue to prefer to keep energy projects out of their community, but want more alternative energy projects nationwide. Hence, it seems unlikely that the network of environmental legislation will be unwound and untangled anytime in the near future.

INFORMAL RULES

One of the most influential informal institutions that drive the regulation of the environment is local preference and culture. As long as citizens desire to prevent the siting of new plants, they will form coalitions and work against these projects. These groups will work to elect local leaders who share their views and in turn will legislate those views. Regulators appointed by the politicians will attempt to uphold that administration's policy objective in order to maintain the progression of their career.

The precedents set by court cases such as Orion v. Salazar illustrate that these environmental regulations may be adjusted at any time. These rules may even be adjusted to such extremes so as to apply to laws retroactively, as was also demonstrated in the BrightSource Energy case. Despite the fact that previous lawsuits and their results can lead some to make informed judgments regarding the expected outcome, analysts cannot guarantee an outcome.

OUTCOMES OF GREEN V. GREEN

Local citizens in collaboration with interest groups will continue to use legal barriers as a roadblock for energy siting. It is low cost of these groups to extend review processes and raise environmental concerns. Local citizens prefer to receive alternative energy when they can free ride when other communities pay the costs and they receive the expected benefits. Conservation groups will continue to push for their policies to be implemented, policies that will be touted when donation season arrives.

Politicians and regulators prefer to continue in their jobs, rather than seek new employment and thus will work to stay in office. National leaders will continue to pass tax incentives for the development of alternative energy. Meanwhile local policymakers will persist in satisfying their constituents' demands and prevent these projects from being sited. Even projects that are as well-funded and nationally supported as the Cape Wind project face significant obstacles in the siting process due to these regulatory barriers.

In each of these cases the outcome is one where local preferences trump global preferences by using environmental institutions. The network of interest groups, litigation, and the culture that has grown up in and around these institutions has shaped the environmental public policy seen today. As national policymakers continue to tout 'green' initiatives, 'green' interest groups will continue to oppose them. The incongruous policy preferences between these two groups will continue to cause lengthy court cases, tedious environmental review processes, and protests. Ultimately it will be our idea of green versus your idea of green.

Appendix
Federal Environmental Regulations
Listed by Agency

BUREAU OF LAND MANAGEMENT

American Indian Religious Freedom Act of 1978—
(Public Law 95–341, 42 U.S.C. 1996 and 1996a)

"It shall be the policy of the United States to protect and preserve for American Indians their inherent right of freedom to believe, express, and exercise the traditional religions of the American Indian, Eskimo, Aleut, and Native Hawaiians."

Archaeological Resource Protection Act of
1979 (ARPA)—(16 U.S.C. 470)

". . .to secure, for the present and future benefit of the American people, the protection of archaeological resources and sites which are on public lands and Indian lands" (Sec. 2(4)(b)).

Bald and Golden Eagle Protection Act:
(16 U.S.C. 668–668d, 54 Stat. 250)

This law provides for the protection of the bald eagle and the golden eagle by prohibiting, except under certain specified conditions, the taking, possession and commerce of such birds. The 1972 amendments increased penalties for violating provisions of the Act or regulations issued pursuant thereto and strengthened other enforcement measures.

The 1978 amendment authorizes the Secretary of the Interior to permit the taking of golden eagle nests that interfere with resource development or recovery operations (see also the Migratory Bird Treaty Act and the Endangered Species Act).

Bankhead-Jones Farm Tenant Act–(7 U.S.C.
1000, 1006, 1010–1012; 50 Stat. 522)

"Directs the Secretary of Agriculture to develop a program of land conservation and utilization in order to correct maladjustments in land use."

Clean Air Act—(42 U.S.C. 7401–7626)

The Clean Air Act is the law that defines EPA's responsibilities for protecting and improving the nation's air quality and the stratospheric ozone layer.

Clean Water Act—(33 U.S.C. 1251—1376)

"The objective of the Federal Water Pollution Control Act, commonly referred to as the Clean Water Act (CWA), is to restore and maintain the chemical, physical, and biological integrity of the nation's waters by preventing pollution, providing assistance to publicly owned treatment works for the improvement of wastewater treatment, and maintaining the integrity of wetlands."

Endangered Species Act (16. 35. 1531–1544)

The law requires federal agencies, in consultation with the U.S. Fish and Wildlife Service and/or the NOAA Fisheries Service, to ensure that actions they authorize, fund, or carry out are not likely to jeopardize the continued existence of any listed species or their habitat. The law also prohibits any action that causes a "taking" of any listed species.

Energy Policy Act of 2005 (42 U.S.C. 15801)

This is the driving initiative for the alternative, or green, energy push. There are portions allocating uses and quality of petroleum, etc. Essentially it is in place to regulate the current energy creators and innovate for more clean energy.

Federal Advisory Committee Act—(5 U.S.C. Appendix 2)

This act is to regulate the amount, longevity, and power of advisory committees. There are so many, with unspecified power, and operating without being in the public that congress found it necessary to regulate these committees and keep their workings open to the public.

Federal Land Policy and Management Act (FLPMA)— (43 U.S.C. 1701 et seq.)

Sets out the need of Federal land to be inventoried, protected, and managed, so as to create an environment to aid domestic growth and protect wildlife and natural resources from overexploitation.

Federal Noxious Weeds Act (title 7. Chapter 41. 2814)

The Act provides for the control and management of non-indigenous weeds that injure or have the potential to injure the interests of agriculture and commerce, wildlife resources, or the public health.

Federal Onshore Oil and Gas Leasing Reform Act of 1987 (FOOGLRA).—(30 U.S.C. § 181 et seq.)

Sets out the criteria for leasing of land for oil and gas harvesting.

Magnuson-Stevens Fishery Conservation and Management Act (16 U.S.C. 1803)

In General any fishery management plan prepared, and any regulation promulgated to implement any such plan, pursuant to this title shall be consistent with the national standards for fishery conservation and management.

Migratory Bird Treaty Act (16 U.S.C. 703–712; Ch. 128; July 13, 1918; 40 Stat. 755)

Establishes a federal prohibition, unless permitted by regulations, to "pursue, hunt, take, capture, kill, attempt to take, capture or kill, possess, offer for sale, sell, offer to purchase, purchase, deliver for shipment, ship, cause to be shipped, deliver for transportation, transport, cause to be transported, carry, or cause to be carried by any means whatever, receive for shipment, transportation or carriage, or export, at any time, or in any manner, any migratory bird, included in the terms of this Convention . . . for the protection of migratory birds . . . or any part, nest, or egg of any such bird."

Mineral Leasing Act (30 U.S.C. 185)

General regulations are as follows

1. Promote to the greatest extent practicable the use of applicant/holder-generated plans of development (PODs).
2. Protect the quality of natural resources and prevent unnecessary environmental damage to lands and resources.
3. Protect the ROW holder's investments in improvements on the ROW.

The objectives of the ROW program are:

A. Principal Use. Recognize that ROWs are a principal or major use of the public lands. As such, the ROW program must receive the level of management interest, priority, and attention commensurate to the other principal or major uses of the public lands and to the magnitude of the impacts associated with the various ROW proposals.
B. Guidance. Provide policy, procedures, and guidance for managing ROWs on public and Federal land so as to:
   1. Coordinate the actions of individuals, governments, and businesses in using lands for ROW purposes.
   2. Minimize the proliferation of separate ROWs.

3. Promote the sharing of ROWs with respect to engineering and technological compatibility, national security, and land use planning.
4. Provide a system of designated ROW corridors on public land to help meet future ROW needs as appropriate.
5. Promote efficiency in granting ROWs.
6. Promote uniform ROW application processing and granting requirements and procedures.

Multiple-Use Sustained-Yield Act—(16 U.S.C. 528 et seq.)

It is the policy of the Congress that the national forests are established and shall be administered for outdoor recreation, range, timber, watershed, and wildlife and fish purposes. The purposes of this Act are declared to be supplemental to, but not in derogation of, the purposes for which the national forests were established as set forth in the Act of June 4, 1897 (16 U.S.C. 475). Nothing herein shall be construed as affecting the jurisdiction or responsibilities of the several States with respect to wildlife and fish on the national forests.

National Environmental Policy Act (NEPA)— ((Pub. L. 91–190, 42 U.S.C. 4321–4347, January 1, 1970, as amended by Pub. L. 94–52, July 3, 1975, Pub. L. 94–83, August 9, 1975, and Pub. L. 97–258, § 4(b), Sept. 13, 1982)

The National Environmental Policy Act (NEPA) requires federal agencies to consider impacts to environment when making decisions and looking at alternatives to any proposed actions on federal land to reduce impact on the environment. Environmental values are to be stressed and taken into consideration. In addition, federal agencies must prepare an Environmental Impact Statement (EIS). EPA reviews and comments on EISs will be prepared by other agencies.

National Historic Preservation Act of 1966 (NHPA)— (16 U.S.C. 470)

This act is in place to protect historic artifacts and places.
Instruction Memorandum No. 2007–097
An act to establish a Program for the Preservation of Additional Historic Properties throughout the Nation.

National Trails System Act of 1968 (16 U.S.C. 1241–51)

In order to provide for the ever-increasing outdoor recreation needs of an expanding population and in order to promote the preservation of, public access to, travel within, and enjoyment and appreciation of the open-air, outdoor areas and historic resources of the Nation, trails should be established (i) primarily, near the urban areas of the Nation, and (ii) secondarily, within scenic areas and along historic travel routes of the Nation which are often more remotely located.

Native American Graves Protection and Repatriation Act of 1990 (NAGPRA)—(25 U.S.C. 3001 et seq.)

This act is in place to protect Native American burial sites, sacred relics, and other such items of import to the culture. "If a discovery occurred in connection with an activity, the person shall cease the activity in the area of the discovery, and make a reasonable effort to protect the items discovered before resuming."

Noise Control Act of 1972, as amended by the Quiet Communities Act of 1978 (42 U.S.C. 4901–4918).

In the past, the EPA coordinated all federal noise control activities. However, in 1981, the Administration at that time concluded that noise issues were best handled at the State or local government level. As a result, the EPA phased out the office's funding in 1982 as part of a shift in federal noise control policy to transfer the primary responsibility of regulating noise to state and local governments. However, the Noise Control Act of 1972 and the Quiet Communities Act of 1978 were not rescinded by Congress and remain in effect today, although essentially unfunded.

Occupational Safety and Health Act (29 U.S.C. 651 et seq.)

See Occupational Safety and Health Act under Bureau of Land Management.

Organic Administration Act— (16 U.S.C. §§ 473–478, 479–482 and 551)

This act, under which most national forests were established, states: "No national forest shall be established, except to improve and protect the forest within the boundaries, or for the purpose of securing favorable conditions of water flows, and to furnish a continuous supply of timber for the use and necessities of citizens of the United States. . ."

Public Rangelands Improvement Act of 1978—(43 U.S.C. 1905)

This Act establishes a national policy to improve the conditions on public rangelands, requires an inventory and consistent federal management policies, and provides funds. It also amends the Wild Free-Roaming Horses and Burros Act and the Federal Land Policy and Management Act of 1976.

Surface Resources Act—(30 U.S.C.611–615; 43 CFR 3715)

The Materials Disposal Act of 1947, 30 U.S.C. secs. 601–604, and the Surface Resources Act of 1955, 30 U.S.C. secs. 601, 603, 611–613, govern the disposal and removal of certain mineral waste materials.

Taylor Grazing Act of 1934—(43 U.S.C. 315)

Intended to "stop injury to the public grazing lands [excluding Alaska] by preventing overgrazing and soil deterioration" (USDI 1988). This Act was preempted by the Federal Land Policy and Management Act of 1976 (FLPMA)."

Water Mitigation Agreement in accordance with MBOGC Order 99–99 Item 6

This Order requires that "The mitigation agreement must provide for prompt supplementation or replacement of water from any natural spring or water well affected by the CBNG project and shall be under such conditions as the parties mutually agree upon."

BUREAU OF INDIAN AFFAIRS

American Indian Religious Freedom Act— (Public Law 95–341, 42 U.S.C. 1996 and 1996a)

See American Indian Religious Freedom Act under Bureau of Land Management.

Mineral Leasing Act—(30 U.S.C. 185)

See Mineral Leasing Act under Bureau of Land Management.

National Environmental Policy Act—(42 U.S.C. 432I)

See National Environmental Policy Act under Bureau of Land Management.

National Historic Preservation Act of 1966 (NHPA)—(16 U.S.C. 470)

See National Historic Preservation Act of 1966 under Bureau of Land Management.

Native American Graves Protection and Repatriation Act of 1990 (NAGPRA)—(25 U.S.C. 3001 et seq.)

See Native American Graves Protection and Repatriation Act of 1990 under Bureau of Land Management.

FEDERAL AVIATION ADMINISTRATION

Federal Aviation Act of 1958

The Act gave the FAA authority to oversee and regulate aviation safety in both military and civilian aircraft. The FAA's initiative is part of a broader effort to try to increase the use of biofuels in the public sector. As a part of a 2011 initiative for cleaner energy production and use, the FAA announced that it was awarding $7.7 million in contracts to aviation companies to promote the use of greener fuels in airplanes.

FISH AND WILDLIFE

Bald and Golden Eagle Protection Act: (16 U.S.C. 668–668d, 54 Stat. 250)

See Bald and Golden Eagle Protection Act under Bureau of Land Management.

Endangered Species Act (16. 35. 1531–1544)

See Endangered Species Act under Bureau of Land Management.

Migratory Bird Treaty Act—(16 U.S.C. 703–712; Ch. 128; July 13, 1918; 40 Stat. 755)

See Migratory Bird Treaty Act under Bureau of Land Management.

National Environmental Policy Act—(42 U.S.C. 432I)

See National Environmental Policy Act under Bureau of Land Management.

National Wildlife Refuge System Improvement Act—(16 U.S.C. 668dd-668ee)

1. The National Wildlife Refuge System is comprised of over 92,000,000 acres of Federal lands that have been incorporated within 509

individual units located in all 50 States and the territories of the United States.

2. The System was created to conserve fish, wildlife, and plants and their habitats.
3. The System serves a pivotal role in the conservation of migratory birds, anadromous and interjurisdictional fish, marine mammals, endangered and threatened species, and the habitats on which these species depend.
4. The System assists in the fulfillment of important international treaty obligations.
5. The System includes lands purchased not only through the use of tax dollars but also through the proceeds from sales of Duck Stamps and national wildlife refuge entrance fees.
6. When managed in accordance with principles of sound fish and wildlife management and administration, fishing, hunting, wildlife observation, and environmental education in national wildlife refuges have been and are expected to continue to be generally compatible uses.

Rivers and Harbors Appropriation Act of 1899—(33 U.S.C. 403)

Section 9 of this Act, commonly known as the Rivers and Harbors Act of 1899, prohibits the construction of any bridge, dam, dike or causeway over or in navigable waterways of the U.S. without Congressional approval. Administration of section 9 has been delegated to the Coast Guard.

FOREST SERVICE

American Recovery and Reinvestment Act—(Public Law 111–5)

On Feb. 13, 2009, Congress passed the American Recovery and Reinvestment Act of 2009 at the urging of President Obama, who signed it into law four days later. A direct response to the economic crisis, the Recovery Act has three immediate goals:

1. Create new jobs and save existing ones
2. Spur economic activity
3. Foster unprecedented levels of accountability in government spending

The Recovery Act intends to achieve those goals by:

1. Providing $288 billion in tax cuts and benefits for millions of working families and businesses

2. Increasing federal funds for education and health care and other programs by $224 billion
3. Making $275 billion available for federal contracts, grants and loans
4. Requiring recipients of Recovery funds to report quarterly on how they are using the money.

Americans with Disabilities Act—(49 U.S.C. § 41705)

The purpose is to

1. Provide a clear and comprehensive national mandate for the elimination of discrimination against individuals with disabilities;
2. Provide clear, strong, consistent, standards
3. Ensure that the Federal Government plays a central role in enforcing the standards; and
4. Invoke the sweep of congressional authority, including the power to enforce the fourteenth amendment and to regulate commerce, in order to address the major areas of discrimination faced day-to-day.

Architectural Barriers Act of 1968—(42 U.S.C. §§ 4151 et seq.)

The ABA requires access to facilities designed, built, altered, or leased with Federal funds. Passed by Congress in 1968, it marks one of the first efforts to ensure access to the built environment. The Access Board develops and maintains accessibility guidelines under this law. These guidelines serve as the basis for the standards used to enforce the law. Four Federal agencies are responsible for these standards: the Department of Defense, the Department of Housing and Urban Development, the General Services Administration, and the U.S. Postal Service.

Bankhead-Jones Farm Tenant Act— (7 U.S.C. 1000, 1006, 1010–1012; 50 Stat. 522)

See Bankhead-Jones Farm Tenant Act under Bureau of Land Management

Bald and Golden Eagle Protection Act: (16 U.S.C. 668–668d, 54 Stat. 250)

See Bald and Golden Eagle Protection Act under Bureau of Land Management.

Clean Water Act—(33 U.S.C. 1251—1376)

See Clean Water Act under Bureau of Land Management.

Comprehensive Environmental Response, Compensation, and Liability Act (CERCLA)—(42 U.S.C. §9601 et seq. (1980))

Comprehensive Environmental Response, Compensation, and Liability Act (CERCLA), commonly known as Superfund, was enacted by Congress on December 11, 1980. This law created a tax on the chemical and petroleum industries and provided broad Federal authority to respond directly to releases or threatened releases of hazardous substances that may endanger public health or the environment.

Electric Consumer Protection Act of 1986— (16 U.S.C.§ 791(a) et seq.)

This is an act to amend the Federal Power Act to provide for more protection to electric consumers.

Endangered Species Act (16. 35. 1531–1544)

See Endangered Species Act under Bureau of Land Management.

Energy Policy Act of 2005 (42 U.S.C. 15801)

See Energy Policy Act of 2005 under Bureau of Land Management.

Farmland Protection Policy Act (Public Law 97–98, 7 U.S.C. 4201)

"The FPPA is intended to minimize the impact Federal programs have on the unnecessary and irreversible conversion of farmland to nonagricultural uses. It assures that—to the extent possible—Federal programs are administered to be compatible with state, local units of government, and private programs and policies to protect farmland. Federal agencies are required to develop and review their policies and procedures to implement the FPPA every two years. The FPPA does not authorize the Federal Government to regulate the use of private or non-federal land or, in any way, affect the property rights of owners."

Federal Lands Recreation Enhancement Act— (16 U.S.C. 6801–6814)

The Federal Lands Recreation Enhancement Act (REA) was enacted December 8, 2004 (REA; PL 108–447, Section 804), and provides Federal land-managing agencies with long-term recreation fee authority. It specifically authorizes these agencies, including the BLM, to reinvest recreation fees at the local recreation sites where they were collected to benefit visitors through enhanced facilities and services.

Federal Land Policy and Management Act of 1976—(43 U.S.C. 1701 et seq.)

See Forest Land Policy and Management Act (FLPMA) of 1976 under Bureau of Land Management.

Federal Power Act—(16 U.S.C. 791–828c; Chapter 285, June 10, 1920; 41 Stat. 1063)

This act deals with the regulation of water resources, hydroelectric generation, electric utility companies dealing with interstate commerce, and public, state, and municipal conservation facilities.

Geothermal Steam Act of 1970 (30 U.S.C. 1004)

This Act governs the lease of geothermal steam and related resources on public lands. The Act prohibits issuing geothermal leases on virtually all U.S. Fish and Wildlife Service-administered lands.

Multiple-Use Sustained-Yield Act—(16 U.S.C. 528 et seq.)

See Multiple-Use Sustained-Yield Act under Bureau of Land Management.

National Environmental Policy Act—(42 U.S.C. 4321)

See National Environmental Policy Act under Bureau of Land Management.

National Forest Management Act—(16 U.S.C. §§ 1600–1614)

The National Forest Management Act reorganized, expanded and otherwise amended the Forest and Rangeland Renewable Resources Planning Act of 1974, which called for the management of renewable resources on national forest lands. It is the primary statute governing the administration of national forests.

National Historic Preservation Act of 1966 (NHPA)— (16 U.S.C. 470)

See National Historic Preservation Act of 1966 under Bureau of Land Management.

Organic Administration Act— (16 U.S.C. §§ 473–478, 479–482 and 551)

See Organic Administration Act under Bureau of Land Management.

Public Rangelands Improvement Act of 1978—(43 U.S.C. 1905)

See Public Rangelands Improvement Act of 1978 under Bureau of Land Management.

Resource Conservation and Recovery Act (Hazardous and Solid Waste Amendments)—(42 U.S.C. § 9601 et seq.)

The Resource Conservation and Recovery Act (RCRA) gives EPA the authority to control hazardous waste from the "cradle-to-grave." This includes the generation, transportation, treatment, storage, and disposal of hazardous waste. RCRA also set forth a framework for the management of non-hazardous solid wastes. The 1986 amendments to RCRA enabled EPA to address environmental problems that could result from underground tanks storing petroleum and other hazardous substances.

Surface Resources Act of 1955—(30 U.S.C.611–615; 43 CFR 3715)

See Surface Resources Act of 1955 under Bureau of Land Management.

Taylor Grazing Act of 1934—(43 U.S.C. 315)

See Taylor Grazing Act of 1934 under Bureau of Land Management.

Toxic Substances Control Act—(15 U.S.C. §2601 et seq.)

The Toxic Substances Control Act of 1976 provides EPA with authority to require reporting, record-keeping and testing requirements, and restrictions relating to chemical substances and/or mixtures. Certain substances are generally excluded from TSCA, including, among others, food, drugs, cosmetics and pesticides.

Wild and Scenic Rivers Act—(16 U.S.C. 1271)

The National Wild and Scenic Rivers System was created by Congress in 1968 (Public Law 90–542; 16 U.S.C. 1271 et seq.) to preserve certain rivers with outstanding natural, cultural, and recreational values in a free-flowing condition for the enjoyment of present and future generations. The Act is notable for safeguarding the special character of these rivers, while also recognizing the potential for their appropriate use and development. It encourages river management that crosses political boundaries and promotes public participation in developing goals for river protection. Each river is administered by either a federal or state agency. Designated segments need not include the entire river and may include tributaries.

Wild Free Roaming Horse and Burro Act of 1971—(16 U.S.C. §§ 1331–1340)

Wild horses and burros are managed, protected, and controlled in accordance with the provisions of the Wild Free-Roaming Horses and Burros Act of 1971, as amended.

TENNESSEE VALLEY AUTHORITY

Tennessee Valley Authority Act—(48 Stat. 58–59, 16 U.S.C. sec. 831)

To improve the navigability and to provide for the flood control of the Tennessee River; to provide for reforestation and the proper use of marginal lands in the Tennessee Valley; to provide for the agricultural and industrial development of said valley; to provide for the national defense by the creation of a corporation for the operation of Government properties at and near Muscle Shoals in the State of Alabama, and for other purposes.

DEPARTMENT OF THE INTERIOR

Public Utility Regulatory Policies act of 1978—(16 U.S.C. 2612)

PURPA was passed in response to the unstable energy climate of the late 1970s. PURPA sought to promote conservation of electric energy. Additionally, PURPA created a new class of nonutility generators, small power producers, from which, along with qualified cogenerators, utilities are required to buy power.

PURPA was in part intended to augment electric utility generation with more efficiently produced electricity and to provide equitable rates to electric consumers. Utility companies are required to buy all electricity from "QFS"—qualifying facilities—at avoided cost. PURPA expanded participation of non utility generators in the electricity market, and demonstrated that electricity from nonutility generators could successfully be integrated with a utility's own supply. PURPA required utilities to buy whatever power is produced by QFS (usually cogeneration or renewable energy.)

NATIONAL PARKS AND RECREATION

Acquired Lands Mineral Leasing Act—(30 U.S.C. 191(b))

This Act authorizes and governs leasing of public lands for developing deposits of coal, phosphates, oil, gas and other hydrocarbons and sodium.

Alaska National Interest Lands Conservation Act—
(16 U.S.C. 3111 et seq.)

This Act designated certain public lands in Alaska as units of the National Park, National Wildlife Refuge, Wild and Scenic Rivers, National Wilderness Preservation and National Forest Systems, resulting in general expansion of all systems.

American Indian Religious Freedom Act—
(Public Law 95–341, 42 U.S.C. 1996 and 1996a)

See American Indian Religious Freedom Act under Bureau of Land Management.

Americans with Disabilities Act—(49 U.S.C. § 41705)

See Americans with Disabilities Act under Forest Service.

Animal Welfare Act—(7 U.S.C., 2131–2159)

The Animal Welfare Act was signed into law in 1966. It is the only Federal law in the United States that regulates the treatment of animals in research, exhibition, transport, and by dealers.

Anti-deficiency Act—(31 U.S.C. 1301)

"The Anti-deficiency Act is one of the major laws through which Congress exercises its constitutional control of the public purse. It evolved over a period of time in response to various abuses."

Antiquities Act of 1906—(16 U.S.C. 431–433)

The act provides protection for any historic or prehistoric ruin or monument, or any object of antiquity, from any person who shall appropriate, excavate, injure, or destroy said objects situated on lands owned or controlled by the Government of the United States, without the permission of the Secretary of the Department of the Government.

Archaeological Resources Protection Act of 1979—
(Public Law 96–95; 16 U.S.C. 470aa-mm)

See Archaeological Resources Protection Act of 1979 under Bureau of Land Management.

Architectural Barriers Act of 1968—(42 U.S.C. §§ 4151 et seq.)

See Architectural Barriers Act of 1968 under Forest Service.

Clean Air Act—(42 U.S.C. §7401 et seq.)

See Clean Air Act under Bureau of Land Management

Coastal Zone Management Act—(Public Law 92–583, 16 U.S.C. 1451–1456)

The Act, administered by NOAA's Office of Ocean and Coastal Resource Management (OCRM), provides for management of the nation's coastal resources, including the Great Lakes, and balances economic development with environmental conservation.

The CZMA outlines two national programs, the National Coastal Zone Management Program and the National Estuarine Research Reserve System. The overall program objectives of CZMA remain balanced to "preserve, protect, develop, and where possible, to restore or enhance the resources of the nation's coastal zone."

Commemorative Works Act—(40 U.S.C. 1001)

The Commemorative Works Act, found at 40 U.S.C. §§8901 et seq., specifies the requirements for development, approval, and location of new memorials and monuments in the District of Columbia and its environs. The Act preserves the urban design legacy of the historic L'Enfant and McMillan Plans by protecting public open space and ensuring that future memorials and monuments in areas administered by the National Park Service and the General Services Administration are appropriately located and designed.

Comprehensive Environmental Response, Compensation, and Liability Act (CERCLA)—(42 U.S.C. §9601 et seq. (1980))

See Comprehensive Environmental Response, Compensation, and Liability Act under Forest Service.

Credit CARD Act of 2009 (Pub. L. 111–24; 123 Stat. 1764)

To amend the Truth in Lending Act to establish fair and transparent practices relating to the extension of credit under an open end consumer credit plan, and for other purposes.

Department of Transportation Act of 1969 (Section 4(f)— Preservation of Parklands)—[42 U.S.C. § 4321]

This section of the Act requires the preservation of publicly owned parklands, waterfowl and wildlife refuges, and significant historic sites. There is a specific finding required for significant publicly owned parklands, recreation areas, wildlife and waterfowl refuges, and all significant historic sites "used" for a highway project. This specific finding requires that 1) the selected alternatives must avoid protected areas, unless not feasible or prudent, and 2) the project includes all possible planning to minimize harm.

Endangered Species Act of 1973—(16 U.S.C. Sections 1531–1544)

See Endangered Species Act under Bureau of Land Management.

Energy Policy Act of 2005 (42 U.S.C. 15801)

See Energy Policy Act of 2005 under Bureau of Land Management.

Equal Employment Opportunity Act of 1972—(42 U.S.C. 2000(e))

The Equal Employment Opportunity Act of 1972 is the act which gives the Equal Employment Opportunity Commission (EEOC) authority to sue in federal courts when it finds reasonable cause to believe that there has been employment discrimination based on race, color, religion, sex, or national origin.

Federal Advisory Committee Act—(5 U.S.C. App.)

See Federal Advisory Committee Act under Bureau of Land Management.

Federal Cave resources Protection Act of 1988— (16 U.S.C. 4301 through 4309)

> "The purposes of this Act are (1) to secure, protect, and preserve significant caves on Federal lands for the perpetual use, enjoyment, and benefit of all people; and (2) to foster increased cooperation and exchange of information between governmental authorities and those who utilize caves located on Federal lands for scientific, educational, or recreational purposes."

Federal Insecticide, Fungicide and Rodenticide Act— (7 U.S.C. sec. 136u)

The objective of FIFRA is to provide federal control of pesticide distribution, sale, and use. All pesticides used in the United States must be registered (licensed) by EPA. Registration assures that pesticides will be properly labeled and that, if used in accordance with specifications, they will not cause unreasonable harm to the environment.

Federal Lands Recreation Enhancement Act- (16 U.S.C. 6801–6814)

See Federal Lands Recreation Enhancement Act under Forest Service.

Federal Managers' Financial Integrity Act of 1982

The Federal Managers' Financial Integrity Act (FMFIA), implemented through the Department's Management Control Program, required all DoD managers to assess the effectiveness of management controls applicable to their responsibilities. If material deficiencies are discovered, managers must report those deficiencies with scheduled milestones leading to the resolution of the deficiency. The establishment of FMFIA material on the web is cost effective because of the world -wide implementation of the program, affecting thousands of individuals throughout the Department.

Federal Water Pollution control Act (Commonly known as the clean water act)—(33 U.S.C. 1251—1376)

See Clean Water Act under Bureau of Land Management.

Freedom of Information Act-(5 U.S.C. § 552)

Like all federal agencies, the Department of Justice (DOJ) generally is required under the Freedom of Information Act (FOIA) to disclose records requested in writing by any person. However, agencies may withhold information pursuant to nine exemptions and three exclusions contained in the statute. The FOIA applies only to federal agencies and does not create a right of access to records held by Congress, the courts, or by state or local government agencies.

General Mining Act of 1872—(30 U.S.C. 29 and 43 CFR)

The General mining Act of 1872 is a federal statute. The Act is one among the major statutes on federal land management policy. The Act generally deals with hard rock mining. The Act authorizes and governs, prospecting and mining for economic minerals, such as gold, platinum, and silver, on federal public lands. The Act promotes the development and settlement of publicly owned lands.

The Act provides that any U.S. citizen of eighteen years and above and any foreign company with subsidiaries incorporated in the U.S can freely enter a public domain land to explore minerals. No permission is required for exploring minerals. Although the Act provides permission to the U.S citizens to explore minerals, exploring in certain public domain lands are excluded by the Act.

Geothermal Steam Act of 1970—(30 U.S.C. 1001–1025)

See Geothermal Steam Act of 1970 under Forest Service.

Government Performance and Results Act of 1993— (39 U.S.C. 2803)

The purposes of this act are as follows:

1. Improve the confidence of the American people in the capability of the Federal Government, by systematically holding Federal agencies accountable for achieving program results;
2. Initiate program performance reform with a series of pilot projects in setting program goals, measuring program performance against those goals, and reporting publicly on their progress;
3. Improve Federal program effectiveness and public accountability by promoting a new focus on results, service quality, and customer satisfaction;
4. Help Federal managers improve service delivery, by requiring that they plan for meeting program objectives and;
5. Improve congressional decision making by providing more objective information on achieving statutory objectives, and
6. Improve internal management of the Federal Government."

Hazardous Materials Transportation Act—(29 U.S.C. 653(b)(1))

The Hazardous Material Transportation Act (HMTA) was published in 1975. Its primary objective is to provide adequate protection against the risks to life and property inherent in the transportation of hazardous material in commerce by improving the regulatory and enforcement authority of the Secretary of Transportation. A hazardous material, as defined by the Secretary of Transportation is, any "particular quantity or form" of a material that "may pose an unreasonable risk to health and safety or property."

Historic Sites, Buildings and Antiquities Act—(16 U.S.C. 461–467)

See Antiquities Act of 1906 under National Parks and Recreation.

Land and Water Conservation Fund Act of 1965— (16 U.S.C. 460d et seq.)

The purposes of this Act are to assist in preserving, developing, and assuring accessibility to all citizens of the United States of America of present and future generations and visitors who are lawfully present within the boundaries of the United States of America such quality and quantity of outdoor recreation resources as may be available and are necessary and desirable for individual active participation in such recreation by (1) providing funds for and authorizing Federal assistance to the States in planning, acquisition, and development of needed land and water areas and facilities and (2) providing funds for the Federal acquisition and development of certain lands and other areas.

Mineral Leasing Act—(30 U.S.C. 185)

See Mineral Leasing Act under Bureau of Land Management.

Mining in the Parks Act—(90 Stat. 1342, 16 U.S.C. § 1901)

The Act addresses mining operations on National Park System lands, authorizing the Secretary of the Interior to promulgate regulations governing activities resulting from the exercise of mining claims within the National Park System.

Museum Act—(20 U.S.C. 80(q) et. seq.)

This act establishes a museum for Native American History. It also makes provisions for materials to be placed in it.

National Environmental Policy Act of 1969—(42 U.S.C. 4321–4347)

See National Environmental Policy Act under Bureau of Land Management.

National Historic Preservation Act of 1966—(16 U.S.C. 470)

See National Historic Preservation Act of 1966 under Bureau of Land Management.

National Park Service Concessions Management Improvement Act of 1998—(16 U.S.C. 5951—5966)

Recognizing the ever-increasing societal pressures being placed upon America's unique natural and cultural resources contained in the National Park System, the Secretary shall continually improve the ability of the National Park Service to provide state- of-the-art management, protection, and interpretation of and research on the resources of the National Park System.

National Park Service organic Act (16 U.S.C. § 1–4)

"An act to establish a National Park Service, and for other purposes. The service thus established shall promote and regulate the use of the Federal areas known as national parks, monuments, and reservations hereinafter specified by such means and measures as conform to the fundamental purposes of the said parks, monuments, and reservations, which purpose is to conserve the scenery and the natural and historic objects and the wildlife therein and to provide for the enjoyment of the same in such manner and by such means as will leave them unimpaired for the enjoyment of future generations."

National Park System General Authorities Act—(16 U.S.C. 6)

The purpose of this act is to include all areas administered by the National Park Service in one National Park System and to clarify the authorities applicable to the system. Areas of the National Park System, the act states,

> "though distinct in character, are united through their inter-related purposes and resources into one national park system as cumulative expressions of a single national heritage."

National Park System Resource Protection Act—(16 U.S.C.19(jj))

The Park System Resource Protection Act allows NPS to seek compensation for injuries to Park System resources and use the recovered funds to restore, replace, or acquire equivalent resources, and to monitor and study such resources. The law initially applied only to marine or Great Lakes resources but was expanded in 1996 as part of the Omnibus Park Act to cover injuries to resources within all National Park System units.

National Parks Air Tour Management Act of 2000 (title VIII of PL 106–181)

Primarily because of concerns that noise from air tours over national parks could impair visitors' experiences and park resources, Congress passed the National Parks Air Tour Management Act of 2000 to regulate air tours. The act requires the Federal Aviation Administration (FAA) and the National Park Service to develop air tour management plans for all parks where air tour operators apply to conduct tours. A plan may establish controls over tours, such as routes, altitudes, time of day restrictions, and/or a maximum number of flights for a given period; or ban all air tours. GAO was asked to (1) determine the status of FAA and the Park Service's implementation of the act; (2) assess how the air tour operators and national parks have been affected by implementation; and (3) identify what issues, if any, need to be addressed to improve implementation.

National Parks Omnibus Management Act of 1998— (16 U.S.C. 3101 et seq.)

The purposes of this title are—(1) to more effectively achieve the mission of the National Park Service; (2) to enhance management and protection of national park resources by providing clear authority and direction for the conduct of scientific study in the National Park System and to use the information gathered for management purposes; (3) to ensure appropriate documentation of resource conditions in the National Park System; (4) to encourage others to use the National Park System for study to the benefit of park management as well as broader scientific value, where such study is consistent with the Act of August 25, 1916 (commonly known as the National Park Service Organic Act, 16 U.S.C. 1 et seq.); and (5) to encourage the publication and dissemination of information derived from studies in the National Park System.

National Trails System Act—(16 U.S.C. 1245)

See National Trails System Act under Bureau of Land Management.

Native American Graves Protection and Repatriation Act of 1990—(25 U.S.C. 3001)

See Native American Graves Protection and Repatriation Act of 1990 under Bureau of Land Management.

Negotiated Rulemaking Act—(5 U.S.C. 561 et seq.)

Its purpose is to establish a framework for the conduct of negotiated rulemaking, consistent with section 553 of this title, to encourage agencies to use the process when it enhances the informal rulemaking process.

Noise Control Act of 1972—(42 U.S.C. 4901 to 4918)

See Noise Control Act of 1972 under Bureau of Land Management.

Occupational Safety and Health Act of 1970—(41 U.S.C. 35 et seq.)

Congress passed the Occupational and Safety Health Act to ensure worker and workplace safety. Their goal was to make sure employers provide their workers a place of employment free from recognized hazards to safety and health.

Oil Pollution Act of 1990—(33 U.S.C. 2701 note)

The Oil Pollution Act (OPA) was signed into law in August 1990, largely in response to rising public concern following the Exxon Valdez incident.

The OPA improved the nation's ability to prevent and respond to oil spills by establishing provisions that expand the federal government's ability, and provide the money and resources necessary, to respond to oil spills.

In addition, the OPA provided new requirements for contingency planning both by government and industry.

Omnibus Consolidated Appropriations Act— (18 U.S.C. § 922(g)(9))

This is a huge act covering many legislative acts and varying sections.

Omnibus Public Land Management Act of 2009 (Pub. L. 111–11; 123 Stat. 991; H.R.)

Designates land in 9 states as potential wilderness areas (Title I). It Grants the state of Alaska a seven-mile easement for constructing a single-lane airport access road through the Izembek National Wildlife Refuge in exchange for the transfer of 43,093 acres of state-owned land to the federal wildlife refuge system. Expands the National Wild and Scenic Rivers System, Adds to the National Trails System, Authorizes $1 million to provide grants to states and Indian tribes to compensate ranchers for livestock killed by wolves and to assist non-lethal methods to reduce wolf attacks, Allows for the preservation of former President William Jefferson Clinton's childhood home as a National Historic Site, and Authorizes appropriations for projects to improve water management and conservation.

Privacy Act of 1974—(5 U.S.C. § 552(a))

"The Privacy Act of 1974, 5 U.S.C. § 552a, establishes a code of fair information practices that governs the collection, maintenance, use, and dissemination of personally identifiable information about individuals that is maintained in systems of records by federal agencies. A system of records is a group of records under the control of an agency from which information is retrieved by the name of the individual or by some identifier assigned to the individual.

The Privacy Act requires that agencies give the public notice of their systems of records by publication in the Federal Register. The Privacy Act prohibits the disclosure of information from a system of records absent the written consent of the subject individual, unless the disclosure is pursuant to one of twelve statutory exceptions.

Rehabilitation Act of 1973—(29 U.S.C. 793(c))

The purpose of the act is to replace the Vocational REHABILITATION ACT, to extend and revise the authorization of grants to States for vocational REHABILITATION services, with special emphasis on services to those with the most severe handicaps, to expand special Federal responsibilities and research and training programs with respect to handicapped individuals, to establish special responsibilities in the Secretary of Health, Education, and Welfare for coordination of all programs with respect to handicapped individuals within the Department of Health, Education, and Welfare, and for other purposes.

Rivers and Harbors Appropriation Act of 1899—33 U.S.C. 403

See Rivers and Harbors Appropriation Act of 1899 under Fish and Wildlife.

Robert T. Stafford disaster Relief and Emergency Assistance Act—(42 U.S.C. 5121–5207)

It is the intent of the Congress, by this Act, to provide an orderly and continuing means of assistance by the Federal Government to State and local governments in carrying out their responsibilities to alleviate the suffering and damage which result from such disasters by–

1. Revising and broadening the scope of existing disaster relief programs;
2. Encouraging the development of comprehensive disaster preparedness and assistance plans, programs, capabilities, and organizations by the States and by local governments;
3. Achieving greater coordination and responsiveness of disaster preparedness and relief programs;
4. Encouraging individuals, States, and local governments to protect themselves by obtaining insurance coverage to supplement or replace governmental assistance;
5. Encouraging hazard mitigation measures to reduce losses from disasters, including development of land use and construction regulations
6. Providing Federal assistance programs for both public and private losses sustained in disasters

Safe Drinking Water Act—(42 U.S.C. § 300(f) et seq.)

The Safe Drinking Water Act (SDWA) is the main federal law that ensures the quality of Americans' drinking water. Under SDWA, EPA sets standards for drinking water quality and oversees the states, localities, and water suppliers who implement those standards. SDWA was originally passed by Congress in 1974 to protect public health by regulating the nation's public drinking water supply. The law was amended in 1986 and 1996 and requires many actions to protect drinking water and its sources: river, lakes, reservoirs, springs, and groundwater wells.

Solid Waste Disposal Act (Includes RCRA)— (42 U.S.C. 6901–6992(k))

The Congress hereby declares it to be the national policy of the United States that, wherever feasible, the generation of hazardous waste is to be reduced or eliminated as expeditiously as possible. Waste that is nevertheless generated should be treated, stored, or disposed of so as to minimize the present and future threat to human health and the environment.

Stevenson-Wydler Technology Innovation Act of 1980— (15 U.S.C. 3701)

It is the purpose of this Act to improve the economic, environmental, and social well being of the United States by

1. Establishing organizations in the executive branch to study and stimulate technology;
2. Promoting technology development through the establishment of cooperative research centers;
3. Stimulating improved utilization of federally funded technology developments, including inventions, software, and training technologies, by State and local governments and the private sector;
4. Providing encouragement for the development of technology through the recognition of individuals and companies which have made outstanding contributions in technology; and
5. Encouraging the exchange of scientific and technical personnel among academia, industry, and Federal laboratories.

Surface Mining Control and Reclamation Act of 1977— (30 U.S.C. 1202)

For the purposes of assisting in the planning and evaluation of reclamation projects pursuant to section 405, and assisting in making the certification referred to in section 411(a), the Secretary shall maintain an inventory of eligible lands and waters pursuant to section 404 which meet the priorities stated in paragraphs (1) and (2) of subsection (a). Under standardized procedures established by the Secretary, States and Indian tribes with approved abandoned mine reclamation programs pursuant to section 405 may offer amendments to update the inventory as it applies to eligible lands and waters under the jurisdiction of such States or tribes.

Telecommunications Act of 1996 (section 704 of PL 104–104)

"The Telecommunications Act of 1996 is the first major overhaul of telecommunications law in almost sixty-two years. The goal of this new law is to let anyone enter any communications business—to let any communications business compete in any market against any other. The Telecommunications Act of 1996 has the potential to change the way we work, live and learn. It will affect telephone service—local and long distance, cable programming and other video services, broadcast services and services provided to schools."

Toxic Substance Control Act—(15 U.S.C. 2601–2692)

See Toxic Substance Control Act under Forest Service.

Volunteers in the Parks Act of 1969—(16 U.S.C. 18(j))

This legislation authorized the Secretary of the Interior to establish a "volunteers in the parks" program to aid in interpretation functions or other visitor services or activities in and related to areas administered by the National Park Service. The legislation provided a vehicle that allowed the Service to utilize volunteer help and services.

Wild and Scenic Rivers Act—(16 U.S.C. 1271)

See Wild and Scenic Rivers Act under Forest Service.

Wilderness Act—(16 U.S. C. 1131–1136)

"Directed the Secretary of the Interior, within ten years, to review every road less area of 5,000 or more acres and every road less island (regardless of size) within National Wildlife Refuge and National Park Systems and to recommend to the President the suitability of each such area or island for inclusion in the National Wilderness Preservation System, with final decisions made by Congress. The Secretary of Agriculture was directed to study and recommend suitable areas in the National Forest System.

The Act provides criteria for determining suitability and establishes restrictions on activities that can be undertaken on a designated area. It authorizes the acceptance of gifts, bequests and contributions in furtherance of the purposes of the Act and requires an annual report at the opening of each session of Congress on the status of the wilderness system.

NATIONAL RESOURCES CONSERVATION SERVICE

Agriculture and Food Act of 1981 (Public Law 97–98)

Commodity program provisions of the Agriculture and Food Act of 1981 are summarized. Price support, loan level, disaster payment, program acreage, and other provisions of the legislation are discussed for wheat, feed grains, cotton, rice, peanuts, soybeans, sugar, dairy, and wool and mohair. The following provisions are also summarized: miscellaneous; grain reserves; the national agricultural cost of production standards review board; agricultural exports and PL-480; food stamps; research, extension, and teaching; resource conservation; credit, rural development, and family farms; and floral research and consumer information.

Commodity Credit Corporation Charter Act as amended (15 U.S.C. 714(c))

The CCC Charter Act, as amended, aids producers through loans, purchases, payments, and other operations, and makes available materials and facilities required in the production and marketing of agricultural commodities.

The CCC Charter Act also authorizes the sale of agricultural commodities to other government agencies and to foreign governments and the donation of food to domestic, foreign, or international relief agencies. CCC also assists in the development of new domestic and foreign markets and marketing facilities for agricultural commodities.

Farm Security and Rural Investment Act of 2002

It is Legislation for conservation funding and for focusing on environmental issues. The conservation provisions help farmers and ranchers meet environmental challenges on their land. This legislation simplifies existing programs and creates new programs to address high priority environmental and production goals. The 2002 Farm Bill enhances the long-term quality of our environment and conservation of our natural resources.

Federal Agriculture Improvement and Reform Act of 1996 (Public Law 104–127)

This report provides an item-by-item description and explanation of the new Act, which will guide agricultural programs from 1996–2000. Signed into law in April, the act makes significant changes in long-standing U.S. agricultural policies. Major changes in U.S. commodity programs are included in the Act's Title I, known as the Agricultural Market Transition Act.

Food, Agriculture, Conservation and Trade Act of 1990 (Public Law 101–624)

The Food, Agriculture, Conservation, and Trade Act of 1990 (P.L. 101–624) establishes a comprehensive framework within which the Secretary of Agriculture will administer agricultural and food programs from 1991 to 1995. This report describes provisions of the 1990 Act as amended by the Omnibus Budget Reconciliation Act of 1990 (P.L. 101–508). Provisions for all major commodity programs, such as income and price support, are reported, as well as general commodity provisions, trade, conservation, research, food stamps, fruit and vegetable marketing, organic food standards, grain quality, credit, rural development, forestry, crop insurance and disaster assistance, and global climate change provisions.

Food Security Act of 1985 as amended (16 U.S.C. 3841 et seq.)

The Food Security Act of 1985 (P.L. 99–198) establishes a comprehensive framework within which the Secretary of Agriculture will administer agriculture and food programs from 1986 through 1990. This report describes the Act's provisions for dairy, wool and mohair, wheat, feed grains, cotton, rice, peanuts, soybeans, and sugar (including income and price supports, disaster payments, and acreage reductions); other general commodity provisions; trade; conservation; credit; research, extension, and teaching; food stamps; and marketing.

Forest and Rangeland Renewable Resources Planning Act of 1974

The purpose of the act is as follows

1. The management of the Nation's renewable resources is highly complex and the uses, demand for, and supply of the various resources are subject to change over time
2. The public interest is served by the Forest Service, Department of Agriculture, in cooperation with other agencies, assessing the Nation's renewable resources, and developing and preparing a national renewable resource and program, which is periodically reviewed and updated
3. To serve the national interest, the renewable resource program must be based on a comprehensive assessment of present and anticipated uses, demand for, and supply of renewable resources from the National Forest Management Act of 1976 Nation's public and private forests and rangelands, through analysis of environmental and economic impacts, coordination of multiple use and sustained yield opportunities as provided in the Multiple-Use, Sustained-Yield Act of 1960 (74 Stat. 215; 16 U.S.C. 528–531), and public participation in the development of the program
4. The new knowledge derived from coordinated public and private research programs will promote a sound technical and ecological base for effective management, use, and protection of the Nation's renewable resources
5. Inasmuch as the majority of the Nation's forests and rangeland is under private, State, and local governmental management and the

Nation's major capacity to produce goods and services is based on these non-federally managed renewable resources, the Federal Government should be a catalyst to encourage and assist these owners in the efficient long-term use and improvement of these lands and their renewable resources consistent with the principles of sustained yield and multiple use

6. The Forest Service, by virtue of its statutory authority for management of the National Forest System, research and cooperative programs, and its role as an agency in the Department of Agriculture, has both a responsibility and an opportunity to be a leader in assuring that the Nation maintains a natural resource conservation posture that will meet the requirements of our people in perpetuity

7. Recycled timber product materials are as much a part of our renewable forest resources as are the trees from which they originally came, and in order to extend our timber and timber fiber resources and reduce pressures for timber production from Federal lands, the Forest Service should expand its research in the use of recycled and waste timber product materials, develop techniques for the substitution of these secondary materials for primary materials, and promote and encourage the use of recycled timber product materials."

Renewable Resources Extension Act of 1978

The Renewable Resources Extension Act provides funding for extension activities related to forestry and natural resources at land grant universities.

Soil and Water Resources Conservation Act of 1977

This Act provides for a continuing appraisal of U.S. soil, water and related resources, including fish and wildlife habitats, and a soil and water conservation program to assist landowners and land users in furthering soil and water conservation.

Soil Conservation and Domestic Allotment Act, as amended (Public Law 74–46, 49 Stat. 163, 16 U.S.C. 590(b)-(f))

This Act authorized the Secretary of Agriculture to provide financial assistance to agricultural producers for carrying out conservation and environmental enhancement measures. It also created a number of Great Plains conservation programs. The 1996 Farm Bill repealed certain of these programs and amended the Act to direct the Secretary to provide financial and technical assistance through the new environmental quality incentives program described in the Food Security Act of 1985 as amended. The Soil Conservation and Domestic Allotment Act continues to authorize other programs.

Notes

NOTES TO CHAPTER 1

1. For a comprehensive list of all the federal regulations we found during our research please see Appendix 1.
2. The reasons we chose to include oil shale will be discussed later in this chapter.

NOTES TO CHAPTER 2

1. Since Hardin first wrote about the commons, scholars have distinguished between common property in which there are no relevant institutions (open access) and common property as a social institution complete with use rights, sanctions, and norms. All references in this discussion to the commons or common property should be understood as open access.
2. Carrying capacity is not a fixed or even constant measure across all systems of grazing. The absolute number of cows able to use a particular pasture without destroying it varies according to timing of the grazing, rest periods, moisture, weather, etc.

References

1964 Wilderness Act. (2008). U.S. Forest Service. Retrieved on February 12, 2012, from http://www.foresthistory.org/ASPNET/policy/Wilderness/1964_Wilderness. aspx.

20% Wind Energy by 2030: Increasing Wind Energy's Contribution to U.S. Electricity Supply. (2008). U.S. Department of Energy. July. Retrieved on January 28, 2012, from www.nrel.gov/docs/fy08osti/41869.pdf.

About Oil Shale. (n.d.). Oil Shale and Tar Sands Information Center. Retrieved on January 30, 2012, from http://ostseis.anl.gov/guide/oilshale/.

About Us. (2009). *Desert Survivors*. Retrieved on February 1, 2012, from http://www.desert-survivors.org/about.htmlhttp://www.desert-survivors.org/about. html.

Advisory Council on Historic Preservation. (2011). Advisory Council on Historic Preservation. October 13. Retrieved on February 15, 2012, from http://www. achp.gov/renewable_energy.html.

Acevedo, Melissa, and Krueger, Joachim, I. (2004). "Evidential reasoning in the Prisoner's Dilemma." *American Journal of Psychology* 118(3): 431–457.

Advantages and Disadvantages Of Geothermal Energy. (n.d.). Renewable & Non-Renewable Energy Sources. Retrieved on February 10, 2012, from http://www. conserve-energy-future.com/Advantages_Disadvantages_GeothermalEnergy.php.

Airborne Wind Turbine. (n.d.). ARPA. Retrieved on February 11, 2012, from http://arpae.energy.gov/ProgramsProjects/OtherProjects/RenewablePower/ AirborneWindTurbine.aspx.

Ajanovic, A. (2011). ScienceDirect -Energy: Biofuels versus food production: Does biofuels production increase food prices?. *Science Direct*. April. Retrieved on December 19, 2011, from www.sciencedirect.com.dist.lib.usu.edu/science/ article/pii/S0360544210002896.

Alexander, C., & Hurt, C. (2007). Biofuels and Their Impact on Food Prices. *Bio-Energy: Fueling America Through Renewable Resources*. September. Retrieved on December 19, 2011, from www.ces.purdue.edu/extmedia/ID/ID-346-W. pdf.

Alternative Energy. (2010). Center for excellence in Photovoltaic research- Making the best of polymer solar cells. Retrieved on March 15, 2011, from http://www. alternative-energy-news.info/best-use-of-polymer-solar-cells/2.

Alternative Fuels and Advanced Vehicles Data Center: Biodiesel Benefits. (n.d.). EERE: Alternative Fuels and Advanced Vehicles Data Center Program. Retrieved on May 29, 2012, from http://www.afdc.energy.gov/afdc/fuels/biodiesel_ben-efits.html.

Alternative Fuels and Advanced Vehicles Data Center: Ethanol. (n.d.). EERE: Alternative Fuels and Advanced Vehicles Data Center Program. Retrieved on February 19, 2012, from http://www.afdc.energy.gov/afdc/ethanol/.

Alternative Fuels and Advanced Vehicles Data Center: Key Federal Legislation. (n.d.). EERE: Alternative Fuels and Advanced Vehicles Data Center Program. Retrieved on May 29, 2012, from http://www.afdc.energy.gov/afdc/laws/key_legislation.

Amar, V. D. (n.d.). "PART 211—Leasing Of Tribal Lands for Mineral Development, Part 211, Chapter I." *Bureau of Indian Affairs, Justia Law*. Retrieved on February 16, 2012, from http://law.justia.com/cfr/title25/25-1.0.1.9.85.html.

American Indian Religious Freedom Act. (n.d.) Federal Historic Preservation Laws. Retrieved on February 4, 2012, from http://www.nps.gov/history/local-law/fhpl_indianrelfreact.pdf.

Anderson, G., and Martin, D. (n.d.). The Public Domain and Nineteenth. *CATO*. Retrieved on February 28, 2012, from www.cato.org/pubs/journal/cj6n3/cj6n3-11.pdf.

Andrews, A. (2011). *Oil Shale: History, Incentives, and Policy*. Darby, PA: DIANE Publishing.

Ankori, M. (2007). Solar Power from Your Large Rooftop Power unto the nations. Israel Business Arena. Retrieved on February 28, 2012, from http://www.solyndra.com/technology-products/200-series/2010.

Appeals Court Quashes Migratory Bird Treaty Act Challenge. (n.d.). American Bird Conservancy. Retrieved on February 17, 2012, from http://www.abcbirds.org/newsandreports/stories/100804.html.

Archaeological Resources Protection Act of 1979. (n.d.). ArchNet. Retrieved February 12, 2012, from http://archnet.asu.edu/topical/crm/usdo.

Arthur, J. D., & Langhus, B. (n.d.). An Overview of Modern Shale Gas Development in the United States. ALL Consulting. Retrieved February 21, 2012, from www.allc.com/publicdownloads/ALLShaleOverv

Bailey, E. (2002). "A Power Struggle: Electric vs. Spiritual." *Los Angeles Times*. July 17. Retrieved on January 9, 2012, from http://articles.latimes.com/2002/jul/17/local/me-medicine17/2.

Baker, M. A., Macris, C., and Patterson, M. H. (2002). "Update on Mining Law and Regulatory Reform." *American Geosciences Institute*. October 21. Retrieved on February 11, 2012, from www.agiweb.org/gap/legis107/mining.html.pdf.

Bartis, J.T., LaTourrette, T., Dixon, L., Peterson, D.J., and Cecchine, G. (2005). "Oil Shale Development in the United States: Prospects and Policy Issues." *RAND Corporation*. MG-414-NETL.

Bauen, A., Gross, R., and Leach, M. (2003). Progress in Renewable Energy. *Environmental International* 29(1): 105–122.

Beerepoot, N., Marmion, A., & Muller, S. (2011). "Renewable Energy: Markets and Prospects by Region." *International Energy Agency*. Retrieved on January 7, 2012, from http://www.iea.org/media/topics/renewables/Renew_Regions.pdf.

Belmans, R., Dragu, C., and Sels, T. (2011). Small Hydro Power: State of the Art and Applications March 14. *Institute of Electrical and Electronic Engineers White paper*. Retrieved on January 28, 2012, from www://wgs.esat.kuleuven.ac.be/electa/IEEE_YRSEPE/papers/18.pdf.

Berger, J. J. (1997). *Charging Ahead: The Business of Renewable Energy and What it Means for America*. New York: Henry Holt & Co.

Biodiesel from Algae. (n.d.) Oilgae. Retrieved on February 29, 2012, from http://www.oilgae.com/algae/oil/biod/biod.html

Blodgett, L., and Slack, K. (2009). "Geothermal 101: Basics of Geothermal Energy Production and Use." *Geothermal Energy Association*. Retrieved on January 11, 2012, from http://smu.edu/smunews/geothermal/Geo101_Final_Feb_15.pdf

Bonsor, K. (n.d.). "How Hydropower Plants Work." *HowStuffWorks*. Retrieved onFebruary 28, 2012, from http://science.howstuffworks.com/environmental/energy/hydropower-plant.htm.

Bozell, J., Moens, L. Petersen, E., Tyson, K. S., and Wallace, R., (2004). "Biomass Oil Analysis: Research Needs and Recommendations." *National Renewable Energy Laboratory*. July 1. Retrieved on December 20, 2011, from http://www1.eere.energy.gov/cleancities/pdfs/bd_status_issues_final.pdf.

BQ-9000. (n.d.). *The National Biodiesel Accreditation Program*. Retrieved November 7, 2011, from http://www.bq-9000.org/

Brian, M. (n.d.). "Diesel Engines vs. Gasoline Engines." *HowStuffWorks*. Retrieved on February 9, 2012, from http://www.howstuffworks.com/diesel1.htm.

BrightSource Energy News Releases (2010). Brightsource Energy. Retrieved on February 23, 2012, from http://ivanpahsolar.com/news-releases.

Brooks, T. (2009). "Biodiesel, LLC." *Western Dubuque*. Retrieved on January 19, 2012, from http://www.wdbiodiesel.net/pdf/WDB%20Fuel%20Matters%20Newsletter%20updated.pdf.

Browner, C. M. (1993). "Pollution Prevention Takes Center Stage." *EPA Journal*. Retrieved on February 15, 2012, from http://www.epa.gov/aboutepa/history/topics/ppa/01.html.

Buchholz, B., Chandler, K., Dinh, H., Fraer, R., McCormick R. L. and Proc, K. (2005). "Operating Experience and Teardown Analysis for Engines Operated on Biodiesel Blends (B20)." *SAE International*. Retrieved on January 19, 2012, from http://www.nrel.gov/vehiclesandfuels/npbf/pdfs/38509.pdf.

Bureau of Reclamation: About Us, Mission. (n.d.). Bureau of Reclamation. Retrieved on February 13, 2012, from http://www.usbr.gov/main/about/mission.html.

Bureau of Reclamation: History. (n.d.). Bureau of Reclamation. Retrieved on February 24, 2012, from http://www.usbr.gov/power/edu/history.html.

Burnett, J. (2009). "Top of Mind: Jim Gordon, Extended Version." *Boston Magazine*. Retrieved on February 8, 2012, from http://www.bostonmagazine.com/articles/jim_gordon/.

Butti, K., and Perlin, J. (2005). "Horace de Saussure and his Hot Boxes of the 1700s." Retrieved on January 20, 2012, from http://solarcooking.org/saussure.html.

Calpine Corporation Company History. (n.d.). FundingUniverse. Retrieved January 5, 2012, from http://www.fundinguniverse.com/company-histories/Calpine-Corporation-company-History.html.

Calpine Corporation. (2010). Business & Company Resource Center. September 9. Retrieved on January 6, 2012, from http://mot5163.wikispaces.com/file/view/Calpin+Company+History.pdf.

Campbell, R. J. (2010). Small Hydro and Low-Head Hydro Power Technologies and Prospects. *Congressional Research Service*. Retrieved on February 27, 2012, from http://nepinstitute.org/get/CRS_Reports/CRS_Energy/Renewable_Fuels/Small_hydro_and_Low-head_hydro_power.pdf.

Cape Wind Threats: The Environment. (n.d.). "Save Our Sound: Alliance to Protect Nantucket Sound." *Cape Wind Threats*. Retrieved on February 8, 2012, from http://www.saveoursound.org/content_item/threats-environment.html.

Cape Wind: America's First Offshore Wind Farm on Nantucket Sound-Frequently Asked Questions. (n.d.) Retrieved on February 8, 2012, from http://www.capewind.org/FAQ-Category4-Cape+Wind+Basics-Parent0-myfaq-yes.htm#16.

Cart, J. (2012). "The power compromise." *Los Angeles Times*. February 5. Retrieved on February 28, 2012, from http://articles.latimes.com/2012/feb/05/local/la-me-solar-desert-20120205.

Carter, J. (Director) (1977). "The President's Proposed Energy Policy." *President's Address*. April 18. Lecture conducted from Carter Administration, Washington D.C.

Century, T. E. (n.d.). "Today in History: September 30." *American Memory from the Library of Congress Home Page*. Retrieved on February 8, 2012, from http://memory.loc.gov/ammem/today/sep30.html.

CERCLA Overview. (2011). US Environmental Protection Agency. December 12. Retrieved on January 14, 2012, from http://www.epa.gov/superfund/policy/cercla.htm.

Citizens' Environmental Assessment on the Decommissioning of Glen Canyon Dam. (2000). Glen Canyon Institute. Retrieved on February 20, 2012 from http://www.glencanyon.org/pdfs/CEA_Report.pdf.

Civil Features Volume IV. (n.d.). *Civil Features.* Retrieved on December 13, 2011, from http://hydropower.inel.gov/techtransfer/pdfs/feasibility_studies_for_small_scale_hydropower_additions-18.pdf.

Clarke, C. (2009). "Ivanpah Solar Electric Generating System." *Basin and Range Watch.* February 15. Retrieved on February 1, 2012, from http://www.basinandrangewatch.org/IvanpahValley.html.

Clarke, S., Courtney, F., Dykes, K., Jodziewicz, L., and Watson, G. (2009). "A Path Forward: A Working Paper of the U.S. Offshore Wind Collaborative." *U.S. Offshore Wind Energy.* October. Retrieved on March 5, 2012, from http://www.usowc.org/pdfs/PathForwardfinal.pdf.

Clean Cities Alternative Fuel Price Report. (n.d.). U.S. Department of Energy. *Energy Efficiency & Renewable Energy.* Retrieved November 7, 2011, from http://www.afdc.energy.gov/afdc/pdfs/afpr_jul_11.pdf.

Coal. (n.d.). *BLM—The Bureau of Land Management.* Retrieved February 11, 2012, from http://www.blm.gov/wo/st/en/prog/energy/coal_and_non-energy.print.html.

Coast Guard, 14 USC § 431: Repealed.

Coastal Zone Management Act of 1972. (2011). National Oceanic Management Act of 1972. March 21. Retrieved on February 5, 2012, from http://coastalmanagement.noaa.gov/about/czma.html#section303.

Code of Federal Regulations (CFR). (2011). U.S. Government Printing Office. March 24. Retrieved on January 13, 2012, from http://www.gpo.gov/fdsys/browse/collectionCfr.action?collectionCode=CFR.

Congressional Action to Help Manage Our Nation's Coasts. (2011, March 22). National Oceanic and Atmospheric Administration. Retrieved on February 5, 2012 from http://coastalmanagement.noaa.gov/czm/czm_act.html.

Corley, A.M. (2010, June). "The Future of Hydropower." *Institute of Electrical and Electronic Engineers Spectrum.* Retrieved on February 28, 2012, from http://spectrum.IEEE.org/energy/renewables/future-of-hydropower.

Country analysis brief: Saudi Arabia. (2010). U.S. Energy Information Administration. July 14. Retrieved on January 16, 2012, from, http://205.254.135.7/countries/country-data.cfm?fips=SA

Current Controversy: Draining Lake Powell. (2002). Kenyon College. Retrieved February 20, 2012 from http://www2.kenyon.edu/projects/Dams/glp04smi.html.

Danelski, D. (2011). "Mojave Desert: First displaced tortoise released." *Press Enterprise.* October 8. Retrieved on February 2, 2012, from http://www.pe.com/local-news/topics/topics-environment-headlines/20111009-mojave-desert-first-displaced-tortoise-released.ece.

Davies, T. (2007). "Ethanol comes with environmental impact, despite green image." *USA Today.* May 5. Retrieved on February 19, 2012, from http://www.usatoday.com/money/industries/environment/2007–05–05-ethanolenvironment_N.htm.

De Paula, M. (2011). "Veggie Oil Smells Better Than Diesel, But It's No Slam Dunk." *Forbes.* May 27. Retrieved on February 7, 2012, from http://www.forbes.com/sites/matthewdepaula/2011/05/27/veggie-oil-might-smell-better-than-diesel-but-its-no-slam-dunk/.

DeCarolis, J. F. and Keith, D. W. (2006). "The economics of large-scale wind power in a carbon constrained world." *Energy Policy* 34(4): 395–410. Retrieved on

January 10, 2012, from http://www.sciencedirect.com/science/article/B6V2W-4D09G8J-2/2/2bea57f41c3e062518fa69e201d3ad0b.

Deichmann, N. (2007). "Seismicity Induced by Water Injection for Geothermal Reservoir Stimulation 5 km Below the City of Basel, Switzerland." *American Geophysical Union*. Retrieved on December 24, 2012, from 2007AGUFM. V53F.08D (Bibcode)

deWitt, P. and deWitt, C. (2008). "How Long Does it Take to Prepare an Environmental Impact Statement." *Environmental Practice* 10(4): 164–174.

Diesel—Overview. (2011). *Yahoo Autos*. Retrieved on December 20, 2011, from http://autos.yahoo.com/green_center-article_110/.

Digest of Federal Resource Laws of Interested to the US Fish and Wildlife Service: National Wildlife Refuge System Administration Act. (n.d.). U.S. Fish and Wildlife Service. Retrieved on January 17, 2012, from http://www.fws.gov/laws/lawsdigest/NWRSACT.HTML.

Dinosaur National Monument—Freemont. (n.d.). U.S. National Park Service. Retrieved on January 30, 2012, from http://www.nps.gov/dino/historyculture/fremont-culture.htm.

Dinosaur National Monument—Viewing Petroglyphs and Pictographs. (n.d.). U.S. National Park Service. Retrieved January 30, 2012, from http://www.nps.gov/dino/historyculture/viewing-petroglyphs-and-pictographs.htm.

Dinosaur National Monument. (n.d.). U.S. National Park Service. Retrieved on January 30, 2012, from http://www.nps.gov/dino/historyculture/douglass.htm.

DOE Awards $12 Million to Spur Rapid Adoption of Solar Energy with the Rooftop Solar Challenge (2011). U.S. Department of Energy. December 1. Retrieved on March 2, 2012, from http://energy.gov/articles/doe-awards-12-million-spur-rapid-adoption-solar-energy-rooftop-solar-challenge.

Dominguez-Escalante Expedition. (n.d.). Utah History To Go. Retrieved on January 30, 2012, from http://historytogo.utah.gov/utah_chapters/trappers,_traders,_and_explorers/dominguez-escalanteexpedition.html\.

The Dominguez-Escalante Expedition. (n.d.). Beehive Archive. Retrieved on January 30, 2012, Retrieved from http://beehivearchive.wordpress.com/2009/02/17/177/.

Douglas, N. C. (1990). *Institutions, Institutional Change and Economic Performance (Political Economy of Institutions and Decisions)*. Cambridge: Cambridge University Press.

Doyle, T. (2006). "Koch's New Fight." *Forbes*. September 21. Retrieved on February 8, 2012, from http://www.forbes.com/2006/09/21/koch-gordon-nantucket-biz_cz_td_06rich400_0921nantucket.html.

Draft Programmatic Environmental Impact Statement for Solar Energy Development in Six Southwestern States. (2010). U.S. Department of Energy. Retrieved on January 27, 2012, from http://solareis.anl.gov/documents/dpeis/Solar_DPEIS_Utah_SEZs.pdf.

DSIRE Glossary. (n.d.). *Database of State Incentive for Renewable Energy*. Retrieved on February 29, 2012, from http://dsireusa.org/glossary/.

Earth Talk: Toxic Substances Control Act of 1976? (2011). *Environmental Magazine*. January 1. Retrieved on January 13, 2012, from http://blastmagazine.com/the-magazine/technology/earth/earthtalk-toxic-substances-control-act-of-1976-toilet-paper-rolls/.

Effect of Federal Safe Drinking Water Act, Clean Water Act and Emergency Planning and Community Right-to-Know Act. (n.d.). New York State Department of Environmental Conservation. Retrieved on February 13, 2012, from http://www.dec.ny.gov/energy/46445.htm.

EIA's Energy in Brief: How much of the world's electricity supply is generated from wind and who are the leading generators?. (2011). U.S. Energy Information

Administration. August 30. Retrieved on May 29, 2012, from, http://www.eia. gov/cfapps/energy_in_brief/wind_power.cfm.

EIA's Energy in Brief: What everyone should know about energy. (2012). *Gloresis Projects: Environmental Issues News & Discussion*. January. Retrieved January 5, 2012, from http://forum.gloresis.com/2012/01/03/eias-energy-in-brief-what-everyone-should-know-abo/.

Eilperin, J. (2012). "Why the Clean Tech Boom Went Bust." *Wired*. January 20. Retrieved on March 1, 2012, from http://www.wired.com/magazine/2012/01/ff_solyndra/all/1.

Energy basics: Photovoltaic cell structures. (2011). U.S. Department of Energy. August 12. Retrieved on February 23, 2012, from http://www.eere.energy.gov/basics/renewable_energy/pv_cell_structures.html.

Energy basics: Photovoltaic electrical contacts and cell coatings. (2011). U.S. Department of Energy. August 12. Retrieved on February 8, 2012, from http://www.eere.energy.gov/basics/renewable_energy/pv_contacts_coatings.html.

Energy basics: polycrystalline thin film used in photovoltaics. (2011). U.S. Department of Energy. August 12. Retrieved on February 23, 2012, from http://www.eere.energy.gov/basics/renewable_energy/polycrystalline_thin_film.html?print.

Energy basics: types of silicon used in photovoltaics. (2011). U.S. Department of Energy. August 12. Retrieved on February 23, 2012, from http://www.eere.energy.gov/basics/renewable_energy/types_silicon.html.

Energy savers. (2011). U.S. Department of Energy. February 9. Retrieved on February 16, 2012, from http://www.energysavers.gov/your_home/electricity/index.cfm/mytopic=10830.

Eller, W., and Krutz, G. (2009). "Policy Process, Scholarship, and the Road Ahead: An Introduction to the 2008 Policy Shootout!" *Policy Studies Journal* 37 (1): 1–4. Retrieved on March 5, 2012, from http://onlinelibrary.wiley.com/doi/10.1111/j.1541-0072.2008.00290.x/abstract.

Endangered Species Act. (2011). U.S. Fish and Wildlife Service. December 8. Retrieved on February 11, 2012, from http://www.fws.gov/endangered/laws-policies/esa-history.html.

Energy Analysis. (2011). National Renewable Energy Laboratory. Retrieved on January 19, 2012, from http://www.nrel.gov/analysis/capfactor.html.

Energy Basics: Wind Power Animation. (2011). U.S. Department of Energy, Energy Efficiency and Renewable Energy. August 12. Retrieved on February 10, 2012, from http://www.eere.energy.gov/basics/renewable_energy/wind_animation_text.html?print.

Energy in Brief: What everyone should know about wind power. (n.d.). U.S. Energy Information & Administration. Retrieved on January 28, 2012, from www.eia.gov/energy_in_brief/wind_power.

Energy Resources. (n.d.). Tribal Energy and Environmental Information Clearinghouse. Retrieved February 8, 2012, from http://teeic.anl.gov/er/index.cfm.

Energy Story: Geothermal Energy. (2011). California Energy Commission. Retrieved January 19, 2012, from http://www.energyquest.ca.gov/story/chapter11.html.

Environmental and Safety Information (2012). Retrieved on February 9, 2012, from www.eap.gov.

Environmental Benefits of Solar Power. (n.d.). *Solar Power*. Retrieved on February 28, 2012, from http://solarpower.com/environmental-advantages/.

Environmental Impacts, Attributes, and Feasibility Criteria. (n.d.). U.S. Department of Energy, Retrieved on 14 Dec. 2011, from www1.eere.energy.gov/geo-thermal/pdfs/egs_chapter_8.pdf.

Environmental Management & Energy News (n.d.). Cape Wind Would Reduce New England Electricity Rates $4.6B Over 25 Years. *Environmental Leader*. Retrieved on March 1, 2012, from http://www.environmentalleader.com/2010/02/11/cape-wind-would-reduce-new-england-electricity-rates-4–6b-over-25-years/.

Environmentalist and Public Participation Era, 1970–1993. (2008). U.S. Forest Service. June 9. Retrieved on January 26, 2012, from http://www.foresthistory.org/ASPNET/Publications/first_century/sec8.htm.

EPA Clean Energy-Environment Guide to Action. (n.d.) *US Environmental Protection Agency*. Retrieved on February 29, 2012, from http://www.epa.gov/statelocalclimate/documents/pdf/guide_action_chap5_s2.pdf.

Eriksson, S., Bernhoff, H., and Leijon, M. (2006). "Evaluation of different turbine concepts for wind power." Renewable and Sustainable Energy Reviews 12 (5): 1419–1434. Retrieved on February 7, 2012, from http://www.sciencedirect.com/science/article/pii/S1364032107000111.

Eskridge, G. E. (2008). "Testimony Before the Subcommittee on National Parks, Forests and Public Lands Committee on Natural Resources U.S. House of Representatives Hearing on 'Paying to Play: Implementation of Fee Authority on Public Lands'." *Natural Resources Committee*. March 7. Retrieved on February 7, 2012, from naturalresources.house.gov/uploadedfiles/eskridgetestimony06.18.08.pdf.

Ethanol Facts and History. (n.d.). *Prodigy Engineering Group*. Retrieved on February 29, 2012, from http://prodigyengr.com/front/showcontent.aspx?fileid=75.

Ethanol History-From Alcohol to Car Fuel (n.d.). Ethonal History. Retrieved on February 29, 2012, from www.ethanolhistory.com.

Ethanol Overview: Industry Growth and Federal Programs. (2011). *Congressional Digest*. 90 (8): 226–230. Retrieved on October 20, 2011, from Academic Search Premier.

European Wind Energy Association, (2010). February. Wind in power: 2009 Statistics. Retrieved February 8, 2012 from: http://ewea.org/fileadmin/ewea_documents/documents/statistics/100401_General_Stats_2009.pdf.

Excessive Endangered Species Act Litigation Threatens Species Recovery, Job Creation and Economic Growth. (2011). *Natural Resources Committee*. December 6. Retrieved on January 14, 2012, from, http://naturalresources.house.gov/News/DocumentSingle.aspx?DocumentID=271408.

Faehner, B. (2006). "The Federal Lands Recreation Enhancement Act: How Recreation Access Fees Are Transforming Public Land Recreation." *Wildlands CPR*. March 3. Retrieved on February 7, 2012, from http://www.wildlandscpr.org/road-riporter/federal-lands-recreation-enhancement-act-how-recreation-access-fees-are-transforming.

Fairley, P. (2009). "China's Potent Wind Potential. Technology Review: The Authority on the Future of Technology." September 14. Retrieved on February 8, 2012, from http://www.technologyreview.com/energy/23460/?a=f.

Fargione, J., Hill, J., Tilman, D., Polasky, S., and Hawthorne, P. (2008). "Land Clearing and the Biofuel Carbon Debt." *Science*. February 29. Retrieved on December 19, 2011, from www.sciencemag.org.dist.lib.usu.edu/content/319/5867/1235.full.pdf?sid=deab8fbd-c294–435c-adc4-e8bb1ebd6f27.

Feasibility Assessment of Water Energy Resources of the United States. (n.d.) Energy Efficiency and Renewable Energy: U.S. Department of Energy. Retrieved on February 26, 2012, from http://hydropower.inl.gov/resourceassessment.

Federal Cave Resources Protection Act of 1988. (n.d.). National Speleological Society. Retrieved on February 6, 2012, from http://www.caves.org/committee/conservation/.

Federal Lands Recreation Enhancement Act (REA). (n.d.). U.S. Fish and Wildlife Service. Retrieved on February 7, 2012, from http://www.fws.gov/laws/lawsdigest/REA.html.

Federal Policy. (n.d.). AWEA. Retrieved on February 11, 2012, from http://www.awea.org/issues/federal_poli.

FERC looks to ease development of small hydropower projects. (n.d.). Federal Energy Regulatory Commission. Retrieved on February 27, 2012, from http://www.ferc.gov/media/news-releases/2010/2010–2/04–15–10-A5.asp#skipnav.

FERC: Off-Limits Sites. (n.d.). Federal Energy Regulatory Commission. Retrieved on February 27, 2012, from http://www.ferc.gov/industries/hydropower/gen-info/licensing/small-low-impact/get-started/sites.asp.

First Tortoise Translocation at Ivanpah. (2011). BrightSource Energy. October 10. Retrieved on February 2, 2012, from http://ivanpahsolar.com/first-tortoise-translocation-at-ivanpah.

Fish and Wildlife Coordination Act. (n.d.). U.S. Fish and Wildlife. Retrieved on February 10, 2012, from http://www.fws.gov/laws/lawsdigest/fwcoord.html.

Fogarty, T., and Lamb, R. (2012). *Investing in the Renewable Power Market: How to Profit from Energy Transformation.* Hoboken, New Jersey: John Wiley & Sons, Inc.

Fraas, L., and Partain, L. (2010). *Solar Cells and Their Applications.* (2nd ed.). Hoboken, New Jersey: John Wiley & Sons, Inc.

Frequently Asked Questions—EERE. (2009). Energy Efficiency and Renewable Energy. October 1–6. Retrieved on May 29, 2012, from http://www1.eere.energy.gov/femp/pdfs/fleetfaq.pdf.

Fuel Economy. (n.d.). Energy Efficiency and Renewable Energy. Retrieved on February 19, 2012, from http://www.fueleconomy.gov.

Fueling Station Locator. (n.d.). EERE: Alternative Fuels and Advanced Vehicles Data Center Program. Retrieved on December 19, 2011, from http://www.afdc.energy.gov/afdc/locator/stations/.

Geophysical Union. (2003). "Dam Removal in the United States: Emerging Needs for Science and Policy." *EOS, Transactions, American Geophysical Union* 84(4), 29–36. Retrieved on January 5, 2012, from http://www.yubashed.org/sites/default/files/null/damrem_doyleetal_2003_damrem.

Geothermal energy technology and current status: An overview. (2002) Science-Direct. Retrieved on February 10, 2012, from http://www.sciencedirect.com/science/article/pii/S1364032102000023.

Geothermal Resources. (n.d.). Tribal Energy and Environmental Information Clearing House. Retrieved on January 6, 2012, from http://teeic.anl.gov/lr/dsp_statute.cfm?topic=12&statute=244.

Geothermal Technologies Program: Mission, Vision, and Goals. (n.d.). U.S. Department of Energy, Energy Efficiency and Renewable Energy. Retrieved February 10, 2012, from http://www1.eere.energy.gov/geothermal/vision_mission_goals.html.

Geothermal Tomorrow. (n.d.). U. S. Department of Energy. Retrieved on February 10, 2012, from www1.eere.energy.gov/geothermal/pdfs/geothermal_tomorrow_2008.pdf.

Glen Canyon Natural History Association. (2010). Glen Canyon Dam. Retrieved on December 21, 2011, from Glen Canyon Natural History Association: http://www.glencanyonnha.org/glen_canyon_dam/glencanyondam.php

Green Trust. (n.d.). *Hydroelectric Power.* Retrieved on December 17, 2011, from Green Trust: http://www.green-trust.org/hydro.htm

Green, M. A., Wang, A., and Wenham, S. R., Zhao, J. (1999). Very High Efficiency Silicon Solar Cells—Science and Technology. IEEE Transactions on Electron Devices 46 (10): 1940–1947. Retrieved on February 22, 2012, from http://IEEExplore.IEEE.org/stamp/stamp.jsp?tp=&arnumber=791982.

Habitat Restoration Means Business. (2011). Salmon Recovery. November. Retrieved on February 28, 2012, from http://www.salmonrecovery.gov/Files/HabitatResMeansBusiness%20111611.pdf.

Hansen, A. C., and Butterfield, C. P. (1993). "Aerodynamics of Horizontal-Axis Wind Turbines." Annual Review of Fluid Mechanics 25: 115–149. Retrieved on February 28, 2012, from http://www.annualreviews.org/doi/pdf/10.1146/annurev.fl.25.010193.000555.

Hansen, L. H., Madsen, P. H., Blaabjerg, F., Christensen, H. C., Lindhard, U., and Iskildsen, K. (2000). "Generators and Power Electronics Technology for Wind Turbines." Paper presented at the 27th Annual Conference of the Institute of Electrical and Electronic Engineers Industrial Electronics Society. Retrieved on January 9, 2012, from http://IEEExplore.IEEE.org/stamp/stamp.jsp?tp=&arnumber=975598.

Hardin, G. (1968). "The Tragedy of the Commons." *Science* 162 (3859): 1243–1248.

Harris, W.R., and Scalan, L. (1909). *The Catholic Church in Utah*. Salt Lake City, UT: Intermountain Catholic Press.

Hau, E. (2006). *Wind Turbines: Fundamentals, Technologies, Application, Economics*. 2nd ed. New York, NY: Springer. Retrieved on January 10, 2012, from http://books.google.com/books?hl=en&lr=&id=tDDGiqfPtKMC&oi=fnd&pg=PR17&dq=technical+approaches+to+producing+wind+power&ots=dlXmdkiEHP&sig=35oEwkdg5vLp-jVXeCDZncafBrU#v=onepage&q&f=false.

Havmøllepark, Nysted. (2010). *DONG Energy*. September 9. Retrieved on February 11, 2012, from http://www.dongenergy.com/nysted/en/Pages/index.aspx.

Hays, S. (1999). The Public Land Question. *Conservation and the Gospel of Efficiency: The Progressive Conservation Movement, 1890–1920*. p. 89–90. Pittsburgh, PA: University of Pittsburgh Press.

Headen, R. C., Bloomfield, S., Warnock, M., and Bell, C. (n.d.) "Property Assessed Clean Energy (PACE) Financing: The Ohio Story." *Bricker & Eckler LLP*. Retrieved on February 29, 2012, from http://pacenow.org/blog/wp-content/uploads/Bricker-Eckler-Ohio-PACE.pdf.

Hickel v. Oil Shale Corp. (1970). 400 U.S. 48, 91 S. Ct. 196, 27 L. Ed. 2D 193.

Hirschman, A. O. (1970). *Exit, Voice, and Loyalty: Responses to Decline in Firms, Organizations, and States*. Cambridge, MA: Harvard University Press.

Historic Sites Act of 1935. (1935). U.S. National Park Service. August 21. Retrieved on February 11, 2012, from www.cr.nps.gov/local-law/FHPL_HistSites.pdf.

History of the Clean Air Act. (2012). U.S. Environmental Protection Agency. February 17. Retrieved on February 23, 2012, from http://epa.gov/air/caa/caa_history.html.

Holm, A., Blodgett, L., Jennejohn, D., and Gawell, K. (2010). "Geothermal Energy: International Market Update." *Geothermal Energy: International Market Update*. Retrieved on Febuary 12, 2012, from www.geo-energy.org/pdf/reports/GEA_International.

How Ethanol is Made. (n.d.). *Renewable Fuels Association*. Retrieved on February 19, 2012, from http://www.ethanolrfa.org/pages/how-ethanol-is-made.

How Geothermal Energy Works. (2009). *Union of Concerned Scientists*. Retrieved on February 10, 2012, from http://www.ucsusa.org/clean_energy/technology_and_impacts/energy_technologies/how-geothermal-energy-works.html.

How Hydroelectric Energy Works. (2006). *Union of Concerned Scientists*. Retrieved on January 11, 2012, from http://www.ucsusa.org/clean_energy/technology_and_impacts/energy_technologies/how-hydroelectrichydrokinetic-energy-works.html.

How Wind Energy Works. (2009). *Union of Concerned Scientists*. December 15. Retrieved on January 19, 2012, from http://www.ucsusa.org/clean_energy/technology_and_impacts/energy_technologies/how-wind-energy-works.html.

Hulen, J. B., Lutz, S. J., Schriener, A. Jr. (2000). "Geothermometry, and Granitoid Intrusions in Well GMF 31–17, Medicine Lake Volcano Geothermal System, California." Twenty-fifth Workshop on Geothermal Reservoir Engineering Stanford University, Stanford, California, January 24–26. SGP-TR-165. Retrieved on May 30, 2012, from http://pangea.stanford.edu/ERE/pdf/IGAstandard/SGW/2000/Lutz.pdf.

Hunt, B. J. (n.d.). *Not Even Past*. Retrieved on February 12, 2012, from http://www.notevenpast.org/texas/rise-and-fall-austin-dam.

Hydraulic Fracturing 101. (n.d.). *Earthworks*. Retrieved onFebruary 21, 2012, from http://www.earthworksaction.org/issues.

Hydroelectric Power: How it works. (n.d.). *USGS Georgia Water Science Center*. Retrieved on February 13, 2012, from http://ga.water.usgs.gov/edu/hyhowworks.html.

Hydroelectric Power: Water power: Micro hydro systems. (n.d.). *Electrovent*. Retrieved on February 27, 2012, from green-trust.org/hydro.htm.

Hydroelectric power and water use. (n.d.) USGS Georgia Water Science Center. Water Science for School. Retreived on May 29, 2012, from http://ga.water.usgs.gov/edu/wuhy.html.

Hydropower basic: Introduction. (n.d.). Microhydro web portal. Retrieved on February 27, 2012, from http://www.microhydropower.net/basics/intro.php#Small.

"Hydropower and the World's Energy Future." (2000). *International Hydropower Association*. Retrieved on December 16, 2011, from http://www.ieahydro.org/reports/Hydrofut.pdf.

Hydro Power Plants. (n.d.). *Bryant University*. Retrieved on May 29, 2012, from web.bryant.edu/~langlois/ems/PowerPlant.html.

Impact of Ethanol Blending. (2008). *National Renewable Energy Laboratory*. November 1. Retrieved on February 19, 2012, from http://www.nrel.gov/analysis/pdfs/44517.pdf.

Industry Statistics. (2011). AWEA. Retrieved on February 11, 2012, from http://www.awea.org/learnabout/industry_stats/index.cfm.

International Rivers. (n.d.). *U.S. Experience*. Retrieved on December 16, 2011, from http://www.internationalrivers.org/files/rrdecompt2.pdf.

It's Elemental-The Element Silicon (n.d.). Science Education at Jefferson Lab. Retrieved on February 28, 2012, from http://education.jlab.org/itselemental/ele014.html.

Ivanpah. (2012). BrightSource Energy. Retrieved on February 2, 2012, from http://www.brightsourceenergy.com/projects/ivanpahhttp://www.brightsourceenergy.com/projects/Ivanpah.

Ivanpah Solar Project Named CSP Project of the year. (n.d.). BrightSource Energy. Retrieved on March 2, 2012, from www.brightsourceenergy.com/images/uploads/press_releases/Ivanpah_CSP_Project_of_the_Year_A.

Kempton, W., and Firestone, J. (2007). "Public Opinion about Large Offshore Wind Power: Underlying Factors." *Energy Policy* 35 (3): 1584–1598.

Kennedy, R. F. (2005). "An Ill Wind Off Cape Cod." *New York Times*. 16 December. Retrieved on March 22, 2012, from http://www.nytimes.com/2005/12/16/opinion/16kennedy.html.

Kimmell, K., and Stalenhoef, D. (2011). "The Cape Wind Offshore Wind Energy Project: A Case Study of the Difficult Transition to Renewable Energy." *Golden Gate University Environmental Law Journal*, 5 (1): 201–225.

Kiser, L. L., and Ostrom, E. (1982). "The Three Worlds of Action: A Metatheoretical Synthesis of Institutional Approaches." *Strategies of Political Inquiry*. Beverly Hills, CA: Sage.

Kowalenko, K. (2011). "Documentary Tracks the Pros and Cons of Wind Energy." *Institute of Electrical and Electronic Engineers*. August 29. Retrieved on February 11, 2012, from http://theinstitute.IEEE.org/technology-focus/technology-topic/documentary-tracks-the-pros-and-cons-of-wind-energyand-power.

Kroon, J. (2009). "ECN: Titanium colored solar cell." *ECN*. December. Retrieved on May 29, 2012, from http://www.ecn.nl/nl/nieuws/newsletter-en/2009/december-2009/titanium-colored-solar-cell/.

Kubiszewski, I. (2008). "Geothermal Steam Act of 1970, United States." *Encyclopedia of Earth*. Retrieved on February 16, 2012, from http://www.eoearth.org/article/Geothermal.

Land & Water Conservation Fund. (n.d.). Iowa Department of Natural Resources. Retrieved on February 8, 2012, from http://www.iowadnr.gov/InsideDNR/GrantsOtherFunding/LandWaterConservationFund.aspx.

Largest offshore wind farm opens off Thanet in Kent. (2010). *British Broadcasting Company*. September 23. Retrieved on February 8, 2012, from http://www.bbc.co.uk/news/uk-england-kent-11395964.

Laws and Regulations Applicable to Geothermal Energy Development. (n.d.). Tribal Energy and Environmental Information Clearinghouse. Retrieved February 16, 2012, from http://teeic.anl.gov/er/geothermal/legal/index.cfm.

Laws and Regulations: Coastal Zone Management Act. (n.d.). Tribal Energy and Environmental Information Clearinghouse. Retrieved on February 13, 2012, from http://teeic.anl.gov/lr/dsp_statute.cfm?topic=7&statute=110.

Laws and Regulations Applicable to Solar Energy Development. (n.d.). Tribal Energy and Environmental Information Clearinghouse. Retrieved on February 13, 2012, from http://teeic.anl.gov/er/solar/legal/index.cfm.

Loan Programs for Renewables. (2011). Database of State Incentives for Renewable Energy. Retrieved on February 29, 2012, from http://www.dsireusa.org/documents/summarymaps/Loan_Map.

Logan, J., and Kaplan, S. (2008). "Wind Power in the United States: Technology, Economic, and Policy Issues." *CRS Report for Congress*. Retrieved on February 8, 2012, from www.fas.org/sgp/crs/misc/RL34546.pdf.

Lund, J. (2004). "100 Years of Geothermal Power Production." *Geo-Heat Center*. September. Retrieved on February 12, 2012, from geoheat.oit.edu/bulletin/bull25–3/art2.pdf.

Lund, J. (2007). "Characteristics, Development and Utilization of Geothermal Resources." *Geo-Heat Center*. June. Retrieved on February 12, 2012, from geoheat.oit.edu/bulletin/bull28–2/art1.pdf.

Luther, L. (2005). "National Environmental Policy Act: Background and Implementation." *CRS Report for Congress*. November 16. Retrieved on January 12, 2012, from http://www.fta.dot.gov/documents/Unit1_01CRSReport.pdf.

LWCF Purchases. (2011). U.S. Forest Service. August 11. Retrieved on February 1, 2012, from http://www.fs.fed.us/land/staff/LWCF/.

Madison, J. (1787). "Federalist No. 10-The Same Subject Continued: The Union as a Safeguard Against Domestic Faction and Insurrection." *New York Packet*. Retrieved on May 29, 2012, from http://www.foundingfathers.info/federalistpapers/fedindex.htm.

Mandil, C., and Shihab-Eldin, A. (2010). "Assessment of Biofuels Potential and Limitations." *International Energy Forum*. February. Retrieved on December 19, 2011, from www.ief.org/PDF%20Downloads/Bio-fuels%20Report.pdf.

Maroney, E. (2011). "Fair seas for Cape Wind." *Barn Stable Patriot*. April 22. Retrieved on February 8, 2012, from http://www.barnstablepatriot.com/home2/index.php?option=com_content&task=view&id=24512&Itemid=30.

Maryland Energy Administration. (n.d.). Maryland Energy Administration. Retrieved on November 7, 2011, from http://energy.maryland.gov/.

May the Peace of the Wilderness Be With You. (n.d.). Colorado Wilderness. Retrieved January 2, 2012, from http://www.coloradowilderness.com/wildpages/dinosaur.html.

Mergel, M. (2009). "Clean Water Act." *Toxipedia*. September 16. Retrieved on February 2, 2012, from http://toxipedia.org/display/toxipedia/Clean+Water+Act.

Mercer-Blackman, V., Samiei, H., and Cheng, K. (2007). "IMF Survey: Biofuel Demand Pushes Up Food Prices". *International Monetary Fund*. October 17. Retrieved on December 19, 2011, from http://www.imf.org/external/pubs/ft/survey/so/2007/RES1017A.htm.

Migratory Bird Treaty Act of 1918. (n.d.). U.S. Fish and Wildlife Service. Retrieved on February 17, 2012, from http://www.fws.gov/laws/lawsdigest/migtrea.html.

Mineral Lands and Mining of 1995, Leases and Prospecting Permits, 30 U.S.C. § 193 (2011).

Mineral Lands and Mining of 1995, Mineral Lands and Regulations in General, 30 U.S.C. § 26 (2011).

Mineral Lands and Mining of 1995, Mineral Lands and Regulations in General, 30 U.S.C. § 28 (2011).

Mineral Lands and Mining, Rights-of-way for pipelines through Federal lands of 1995, 30 U.S.C. 185 (2011).

Mineral Leasing Act of 1920 as Amended, 30 U.S.C. § 181 (2007).

Mines and Mining. "Mining Claims; Affidavit of performance of annual labor or payment of maintenance fee." Title 40 Utah Code Annotated, 1–6.

Moe, T. M. (1988). The Organization of Interests: Incentives and the Internal Dynamics of Political Interest Groups. Chicago, IL: Chicago University Press.

Mojave Desert Blog (2010). Mojave Desert Blog. January 1. Retrieved on February 1, 2012, from http://www.mojavedesertblog.com/search?updated.

Morris, L. (2011). "Small Hydro Gets BIG." *PennEnergy*. March 14. Retrieved on January 12, 2012, from http://www.pennenergy.com/index/power/display/9549012947/articlespower-engineering/volume-114/issue-11/features/small-scale-hydro-getsbig.html.

Nanosolar. (2011). Production Process, Nanosolar process. Retrieved on February 24, 2012, from http://www.nanosolar.com/technology/production-process.

National Biodiesel Board. (2011). *National Biodiesel Board*. Retrieved on November 7, 2011, from http://www.biodiesel.org.

National Environmental Policy Act. (n.d.). U.S. Environmental Protection Agency. Retrieved February 21, 2012, from http://www.epa.gov/compliance/nepa/.

National Environmental Policy Act. (2011). U.S. Environmental Protection Agency. Retrieved on January 14, 2012, from http://www.epa.gov/region1/nepa/.

National Forest Management Act of 1976. (n.d.). Federal Wildlife Laws Handbook. Retrieved on February 7, 2012, from http://www.thecre.com/fedlaw/legal14/nfma.htm.

The National Grid Company. (2001). "Reactive Power." *Market Development 1–3*. Retrieved on January 11, 2012, from http://www.nationalgrid.com/NR/rdonlyres/43892106–1CC7–4BEF-A434–7359F155092B/3543/Reactive_Introduction_oct01.pdf.

National Renewable Energy Laboratory. (n.d.). *National Renewable Energy Laboratory*. Retrieved on February 19, 2012, from http://www.nrel.gov.

National Wildlife Refuge System Administration Act of 1966, 16 U.S.C. § 668DD-668EE (2010).

Navarro, J. (2009). "Green Scene: Solar Power Dreamin'." *Defenders*. Fall. Retrieved on February 13, 2012, from http://www.defenders.org/newsroom/defenders_magazine/fall_2009/green_scene_solar_power_dreamin.php.

Neal, R. (2009). Storm Over Mass. Windmill Plan. *CBS News*. February 11. Retrieved on February 8, 2012, from http://www.cbsnews.com/stories/2003/06/26/sunday/main560595.shtml.

Ninth Circuit Reverses Lower Court Ruling and Halts Development on 10,000 Year Old Sacred site at Medicine Lake. (2006). *Stanford Law School News*

Center. November 9. Retrieved on January 20, 2012, from http://www.law.stanford.edu/news/pr/45/.

Noise Control Act of 1972. (1996). The Bureau of National Affairs, Inc. Retrieved on February 7, 2012, from http://www.epa.gov/air/noise/noise_control_act_of_1972.pdf.

NPS Archaeology Program: Curation of Federally-owned and administered archaeological collections, 36 CFR 79. (2011).

Obama, B. The White House, Office of the Press Secretary. (2011). *Remarks by the president in state of union address.* Retrieved on January 5, 2012, from http://www.whitehouse.gov/the-press-office/2011/01/25/remarks-president-state-union-address.

Oil Shale. (n.d.). ACS: Public Outreach and Education. Retrieved on January 30, 2012, from www.ems.psu.edu/~pisupati/ACSOutreach/Oil_Shale.html.

Olson, M. (1965). *The logic of collective action; public goods and the theory of groups.* Cambridge, MA: Harvard University Press.

Oregon State Government. (2012). *Oregon Department Of Energy: Renewable Energy Small/Low-Impact Hydropower.* Retrieved on February 12, 2012, from http://www.oregon.gov/ENERGY/RENEW/Hydro/Hydro_index.shtml.

Origins of the Fish and Wildlife Service. (2009). US Fish and Wildlife Service. Retrieved on December 20, 2012, from http://training.fws.gov/history/TimelinesOrigins.html.

Orion v. Norton, Civil Action No. 04–0791 (RCL) (2006).

Ostrom, E. (1996). "Institutional Rational Choice: An Assessment of the IAD Framework." Presented at the 1996 Annual Meetings of the American Political Science Association, San Francisco, August 29-September 1, 1996.

Ostrom, E. (1999). "Institutional Rational Choice: An Assessment of the Institutional Analysis and Development Framework." In P.A. Sabatier (ed.), *Theories of the Policy Process*, pp. 35–72. Boulder, CO: Westview Press.

Ostrom, E. (2007). "Institutional Rational Choice: An Assessment of the Institutional Analysis and Development Framework." In P. A. Sabatier (ed.), *Theories of the Policy Process*, 2nd ed., pp. 21–64. Boulder, CO: Westview Press.

Ostrom, E. (2011). "Background on the Institutional Analysis and Development Framework." *Policy Studies Journal* 39 (1): 7–27. Retrieved on February 22, 2012, from 10.1111/j.1541–0072.2010.00394.x.

Overview- The Clean Air Act Amendments of 1990. (n.d.) Environmental Protection Agency. Retrieved on May 29, 2012, from http://epa.gov/oar/caa/caaa_overview.html.

Painuly, J. (2001). "Barriers to renewable energy penetration; a framework for analysis." *Renewable Energy* 24 (1): 73–89. Retrieved on March 5, 2012, from http://www.sciencedirect.com/science/article/pii/S0960148100001865.

Paish, O. (2002). "Small Hydro Power: Technology and Current Status." *Renewable and Sustainable Energy Reviews* 6 (6): 537–556.

Part 3200—Geothermal Resource Leasing. (n.d.) Part 3200, Chapter II. Bureau of Land Management. *Justia Law.* Retrieved on February 16, 2012, from http://law.justia.com/cfr/title43/43–2.1.1.3.53.html#43:2.1.1.3.53.1.

Photovoltaic Effect, The. (n.d.). The EncycloBEAMia. Retrieved on March 1, 2012, from http://encyclobeamia.solarbotics.net/ar.

Pit River Tribe v. U.S. Forest Service. 469 F.3d 768 (2006). Lewis and Clark Law School's Environmental Law Online. Retrieved on January 11, 2012, from http://www.elawreview.org/summaries/natural_resources/national_forest_management_act/pit_river_tribe_v_us_forest_se.html#_edn1.

Pit River Tribe v. U.S. Forest Service. 469 F.3d 768 (2006). Retrieved January 19, 2012, from http://ftp.resource.org/courts.gov/c/F3/469/469.F3d.768.04–15746.html.

Pit River Tribe. (2011). Letter to Catarina de Albuquerque. February 24. Retrieved on January 19, 2012 from, http://www.treatycouncil.org/PDF/Image022811103647.Water%20&%20Sanitation%20Expert.2011.final.pdf.

Polski, M. and Ostrom, E. (1999). "An Institutional Framework for Policy Analysis and Design. Workshop in Political Theory and Policy Analysis Working Paper W98–27." *George Mason University*. Retrieved on March 5, 2012 from http://mason.gmu.edu/~mpolski/documents/PolskiOstromIAD.pdf.

Power Benefits of the Lower Snake River Dams. (2009). Bonneville Power Administration. Retrieved January 19, 2012, from http://www.bpa.gov/corporate/pubs/fact_sheets/09fs/Fact_Sheet_-_Power_benefits_of_the_lower_Snake_River_dams.pdf.

Proficiency Testing. (2012). American Society for Testing and Materials. Retrieved on March 3, 2012, from http://www.astm.org/STATQA/biodiesel.htm.

Property Assessed Clean Energy (PACE). 2012. *Database of State Incentives for Rewnewable Energy*. Retrieved February 29, 2012, from http://www.dsireusa.org/documents/summarymaps/PACE_Financing_Map.pdf.

Protecting Religious Freedom and Sacred Sites. (2008). Friends Committee on National Legislation. March 17. Retrieved on January 7, 2012, from http://fcnl.org/issues/nativeam/protecting_religious_freedom_and_sacred_sites/.

Protection of Environment, 40 CFR 1502, (2005).

Protection of Environment 40 CFR § 1501.2 (2005).

Protection of Environment, 40 CFR § 1508.4. (2005).

Protection of Environment, 40 CFR § 1508.9 (2005).

Public Lands: Interior, 43 CFR 10, (2012).

Public Lands: Interior, Mineral Management 43 C.F.R. Part 9, § 3851.3(b) (1993).

Public Land Ownership by State. (n.d.) National Wilderness Institute. Retrieved on March 5, 2012, from http://www.nrcm.org/documents/publiclandownership.pdf.

Ray, R. (2010). "Regulating Small Hydro." *Hydro Review*. April 20. Retrieved on January 9, 2012, from http://www.renewableenergyworld.com/rea/news/article/2010/04/reguating-small-hydro.

Reisner, M. (1993). *Cadillac Desert: the American West and its disappearing water*. New York, NY: Penguin.

Renewable Electricity Production Tax Credit. (2011). *Database of State Incentive for Renewable Energy*. Retrieved February 29, 2012, from http://dsireusa.org/incentives/incentive.cfm?Incentive_Code=US13F.

Renewable energy sources in the United States. (2011). National Atlas. January 26. Retrieved on February 24, 2012, from http://nationalatlas.gov/articles/people/a_energy.html.

Renewable Energy Sources in the United States. (2011). National Atlas. January 26. Retrieved on February 24, 2012, from http://nationalatlas.gov/articles/people/a_energy.htmlhttp://nationalatlas.gov/articles/people/a_energy.html.

Ring, R. (2010). Wilderness Act. *Red Lodge Clearinghouse*. Retrieved on February 13, 2012, from http://rlch.org/content/wilderness-act.

Robert, B. L. Jr. (1997). "Renewable Energy: Not Cheap, Not 'Green.'" *CATO Institute*. Retrieved on February 13, 2012, from http://www.cato.org/pubs/pas/pa-28.

Roberts, J. (2011). "Energy In America: Oil Shale 'Boom Or Bust?'. *Fox News*. Retrieved on January 30, 2012, from http://www.foxnews.com/us/2011/07/13/energy-in-america-oil-shale-boom-or-bust/.

Rogers, J. (2006). "Glen Canyon Unit." *Bureau of Reclamation*. Retrieved on December 16 2011, from http://www.usbr.gov/projects//ImageServer?imgName=Doc_1232657383034.pdf.

Rogers, S. M. (2009). "History of Litigation Concerning Hydraulic Fracturing to Produce Coalbed Methane." *Interstate Oil & Gas Company Commission.* January 1. Retrieved on February 8, 2012, from www.iogcc.state.ok.us/Websites/iogcc/Images/Marvin%20Rogers%20Paper%20of%20History%20of%20LEAF%20Case%20Jan.%202009.pdf.

Romm, J. (2011). EIA: Renewable resources delivered 11% of U.S. energy production in 2010, just like nuclear power. *ThinkProgress.* April 11. Retrieved on March 5, 2012, from http://thinkprogress.org/romm/2011/04/12/207876/eia-renewable-energy-nuclear-power/.

Rosenthal, E. (2011). "Green Development? Not in My (Liberal) Backyard." *New York Times.* March 12. Retrieved on February 8, 2012, from http://www.nytimes.com/2011/03/13/weekinreview/13nimby.html?pagewanted=all.

Ross, A. (2010). "22 Largest Bankruptcies in World History." *InstantShift.* February 3. Retrieved on February 29, 2012, from http://www.instantshift.com/2010/02/03/22-largest-bankruptcies-in-world-history/.

Rossomando, J. (2011). "Are Environmentalists an Obstacle to Clean Energy Production." *Daily Caller.* March 28. Retrieved on February 13, 2012, from http://dailycaller.com/2011/03/28/are-environmentalists-an-obstacle-to-clean-energy-production/.

Rothenberg, L. S. (1992). *Linking Citizens to Government: Interest Group Politics at Common Cause.* New York, NY: Cambridge University Press.

Rothstein, D. S. (2006). "Selected Federal Wildlife and Environmental Laws Applicable to Wind Energy Development." *U.S. Fish and Wildlife Service.* June 6. Retrieved on February 13, 2012, from www.fws.gov/midwest/greatlakes/windpowerpresentations/rothstein.pdf.

RPS Policies. (2012). Database of State Incentives for Renewable Energy. *DSIRE.* Retrieved on February 29, 2012, from http://www.dsireusa.org/documents/summarymaps/RPS_map.pdf.

Rybach, L. (2007). Geothermal Sustainability. *Geo-Heat Center: Oregon Institute of Technology.* Retrieved on February 12, 2012, from geoheat.oit.edu/bulletin/bull28-3/art2.pdf.

Safe Drinking Water Act (SDWA). (n.d.). U.S. Environmental Protection Agency. Retrieved on February 8, 2012, from http://water.epa.gov/lawsregs/rulesregs.

San Diego Natural History Museum Fossil Mysteries: Geologic Timeline. (n.d.). *San Diego Natural History Museum.* Retrieved on January 30, 2012, from http://www.sdnhm.org/archive/exhibits/mystery/fg_timeline.html.

San Juan River Basin. (2007). The Colorado Pikeminnow and the Razorback Sucker. *San Juan River Basin Recovery Implementation Program.* Retrieved on January 2, 2012, from http://www.fws.gov/southwest/sjrip/GB_FS.cfm.

Sanchez, E., Torres, C. F., Guillen, P., and Larrazabal, G. (2011). Modeling and simulation of the production process of electrical energy in a geothermal power plant. *Mathematical and Computer Modeling 17 (2):* 1247–1249.

Saudi Arabia. (2010). Energy Information Administration. June 30. Retrieved on January 31, 2012, from http://www.eia.gov/countries/country-data.cfm?fips=SA.

Schmalensee, R. (2010). "Renewable Electricity Generation in the United States" in *Harnessing Renewable Energy in Electric Power Systems: Theory, Practice, Policy.* Moselle, B., Padilla, J., and Schmalensee, R. (eds). Washington, D.C.: Earthscan.

Schwartz, F. H. and Shahidehpour, M. (n.d.). "Small Hydro as Green Power." *Institute for Electrical and Electronic Engineers.* Retrieved on March 14, 2011, from http://IEEExplore.IEEE.org/xpls/abs_all.jsp?arnumber=4057327&tag=1.

S.D. Warren Co v. Maine Board of Environmental Protection. (2011). OYEZ. Retrieved on February 4, 2012, from http://oyez.org/cases/2000–2009/2005/2005_04_1527#sort=ideology.

Seelye, K. Q. (2010). "Regulators Approve First Offshore Wind Farm in U.S." *New York Times*. April 28. Retrieved on March 22, 2011, from http://www.nytimes.com/2010/04/29/us/29wind.html?_r=3&src=mv&ref=general.

Seltenrich, N. (2011). "Embattled Ivanpah Solar Plant Endangers Tortoises, Gets Brown's Approval." *East Bay Express*. July 26. Retrieved on December 21, 2011, from http://www.eastbayexpress.com/92510/archives/2011/07/26/embattled-ivanpah-solar-plant-endangers-tortoises-gets-browns-approval.

Service Issues Biological Opinion for Ivanpah Solar Electric Project; BLM Lifts Suspension of Activities Order. (2011). U.S. Fish and Wildlife. June 10. Retrieved on February 15, 2012, from http://www.fws.gov/cno/press/release.cfm?rid=239.

Shale Gas Timeline. (2011). BreakThrough. December 20. Retreived on January 15, 2012, from http://thebreakthrough.org/blog/Shale_Gas_Timeline_2.png.

Sharp, C. A. (n.d.). "Exhaust Emissions and Performance of Diesel Engines with Biodiesel Fuels." *National Biodiesel Board*. Retrieved on November 7, 2011, from www.biodiesel.org/resources/reportsdatabase/reports/gen/19980701_gen-065.pdf.

Shaun, G. (2010). *Mojave Desert Blog*. January 1. Retrieved on February 1, 2012, from http://www.mojavedesertblog.com/search?updated.

Sierra Club v. Marita, 46 F.3d 606 (7th Cir. 1995). *Invispress: Law School Case Briefs*. Retrieved on January 27, 2012, from http://www.invispress.com/law/natural/marita.html.

Smajgl, A., Leitch, A. Lynam, T. (eds.). (2009). "Outback Institutions: An application of the Institutional Analysis and Development (AID) framework to our case studies in Australia's outback." *Desert Knowledge CRC Report* 31: 1–65. Retrieved on November 12, 2011, from http://www.desertknowledgecrc.com.au/resource/DKCRC-Report-31-Outback-Institutions_Application-of-the-IAD-framework.pdf.

Solar Energy Development Draft Programmatic EIS. (n.d.). *Solar Energy Development Programmatic EIS Information Center*. Retrieved on February 24, 2012, from http://solareis.anl.gov/documents/dpeis/index.cfm#vol1.

Solar Energy Development Draft Program Environmental Impact Statement. (n.d.) *Solar Energy Development Program*. Retrieved on March 5, 2012, from http://solareis.anl.gov/documents/dpeis/index.cfm.

Solyndra LLC. (2012). Solyndra greenhouse: solyndra agricultural solar products. Retrieved on February 28, 2012, from http://photovoltaics.sandia.gov/docs/PVwww.solyndra.com/technology-products/greenhouse/.

Stojmirovic, G., and Chu, T. (n.d.). "A Practical Guide to Assessment and Implementation of Small Hydropower." *International Network on Small Hydro Power*. Retrieved on December 15, 2011, from www.inshp.org/THE%20 3rd%20HYDRO%20POWER%20FOR%20TODAY%20Forum/Presentations/Australia/A%20Practical%20uide%20to%20Assessment%20and%20 Implementation%20of%20Sm%20Hydropower.pdf.

Stanford Law School. (n.d.). *Environmental law clinic*. Retrieved on January 19, 2012, from http://www.law.stanford.edu/program/clinics/environmental/.

Summary of the Clean Air Act. (2010). U.S. Environmental Protection Agency. May 18. Retrieved on February 13, 2012, from http://www.epa.gov/lawsregs/laws/caa/index.html.

Summary of the Endangered Species Act. (2011) *U.S. Environmental Protection Agency*. August 11. Retrieved on January 12, 2012, from http://www.epa.gov/lawsregs/laws/esa.html.

Summary of the Pollution Prevention Act: 42 U.S.C. §13101. (n.d.). US Environmental Protection Agency. Retrieved on February 24, 2012, from http://www.epa.gov/lawsregs/laws/ppa.html.

Summary of the Resource Conservation and Recovery Act: 42 U.S.C. § 690. (n.d.). US Environmental Policy Agency. Retrieved on February 24, 2012, from http://www.epa.gov/lawsregs/laws/rcra.html.

Summary of the Toxic Substances Control Act: 15 U.S.C. § 2601 (1976). *U.S. Environmental Protection Agency.* Retrieved on February 24, 2012, from http://www.epa.gov/lawsregs/laws/tsca.html.

Tar Sands Basics. (n.d.). *Oil Shale and Tar Sands Information Center.* Retrieved on February 28, 2012, from http://ostseis.anl.gov/guide/tarsands/index.

Tester, J. W., Anderson B., Batchelor, A., Blackwell, D., DiPippo, R., Drake, E., Garnish, J., Livesay, B., Moore, M., Nichols, K., Petty, S., Toksoz, M., and Veatch, R. (2006). "The Future of Geothermal Energy." *Massachusetts Institute of Technology.* Retrieved on February 10, 2012, from geothermal.inel.gov/publications/future_of_geothermal_energy.pdf.

The history of solar. (n.d.). U.S. Department of Energy. Retrieved on February 22, 2012, from www1.eere.energy.gov/solar/pdfs/solar_timeline.pdf.

The Photovoltaic Effect. (n.d). Sandia National Laboratories. Retrieved on February 28, 2012, from http://photovoltaics.sandia.gov/docs/PVFEffIntroduction.htm.

The World's First Hydroelectric Power Plant Began Operation. (n.d.). *America's Story from America's Library.* Retrieved on February 8, 2012, from http://www.americaslibrary.gov/jb/gilded/jb_gilded_hydro_2.html.

Title 30,181. Lands subject to disposition; persons entitled to benefits; reciprocal privileges; helium rights reserved. (n.d.). *Legal Information Institute.* Retrieved on February 11, 2012, from http://www.law.cornell.edu/U.S.C.ode/30/181.html.

Tom Miller Dam and Lake Austin. (n.d.). *Lower Colorado River Authority.* Retrieved on February 12, 2012, from http://www.lcra.org/water/dams/miller.html.

Trembath, A. (2011, December 20). History of the Shale Gas Revolution. TheBreak-Through Institute. Retrieved January 9, 2012, fromhttp://thebreakthrough.org/blog/2011/12/history_of_the_shale_gas_revolution.shtml

True Grit: Fundamental Misunderstandings—The Truth about the Land and Water Conservation Fund. (2007). *Wilderness Society.* June 27. Retrieved on February 8, 2012, from http://wilderness.org/content/true-grit-fundamental-misunderstandings-truth-about-land-and-water-conservation-fund.

TVA: About TVA. (n.d.). *Tennessee Valley Authority.* Accessed February 13, 2012, Retrieved on from http://www.tva.com/abouttva/index.htm.

TVA: From the New Deal to a New Century. (n.d.). *Tennessee Valley Authority.* Retrieved on May 29, 2012, from http://www.tva.com/abouttva/history.htm.

Types of Solar Power. (n.d.). Find Solar Panel Installers & Solar Installation Pros. Retrieved on February 24, 2012, from http://www.getsolar.com/why_solar_types-of-solar-power.php.

Types of Wind Turbines. (2011). U.S. Energy Information Agency. Retrieved on January 19, 2012, from http://www.eia.gov/energyexplained/index.cfm?page=wind_types_of_turbines.

Uintah County Profile. (n.d.). Pioneer: Utah's Online Library. Retrieved on January 30, 2012, from http://pioneer.utah.gov/research/utah_counties/uintah.html.

U.S. Department of Agriculture Factsheet. (n.d.). National Biodiesel Board. Retrieved on November 7, 2011, from www.biodiesel.org/usda/pdfs/USDA_Ed_Factsheet2010.

U.S. Department of Agriculture Rural Development Program. (n.d.). U.S. Department of Agriculture. Retrieved on November 7, 2011, from www.rurdev.usda.gov.

United States v. Midwest Oil Co., 236 U.S. 459 (1915).

Urbanchuk, J. (2010). "Current State of the U.S. Ethanol Industry". *Cardno ENTRIX*. November 30. Retrieved on May 29, 2012 from http://www1.eere. energy.gov/biomass/pdfs/current_state_of_the_us_ethanol_industry.pdf.

Utah's Prehistory in a Nutshell. (n.d.). Utah State History. Retrieved on January 30, 2012, from http://history.utah.gov/archaeology/i_love_archaeology/utah's_ prehistory.html.

Ute Indians. (n.d.). Utah History to Go. Retrieved on January 30, 2012, from http://historytogo.utah.gov/utah_chapters/american_indians/uteindians. html.

Utilities and Wind Power. (n.d.). American Wind Energy Association. Retrieved on January 28, 2012, from http://www.awea.org/learnabout/utility/index.cfm?C FID=149852473&CFTOKEN=83820250&jsessionid=7830a4bcdf30d6b8d59 46534b18303330104.

Vulcan Street Plant. (n.d). American Society of Mechanical Engineers. Retrieved on February 10, 2012, from http://files.asme.org/ASMEORG/Communities/ History/Landmarks/5657.pdf.

Wang, M., Wu, M., and Huo, H. (2007). "Life-cycle energy and greenhouse gas emission impacts of different corn ethanol plant types." *Environmental Research Letters* 2 (2).

Water Pollution Control Act of 1972, 33 U.S.C. § 26, (2012).

Water Power Program: History of Hydropower. (n.d.). U.S. Department of Energy, Energy Efficiency and Renewable Energy. Retrieved on February 8, 2012, from http://www1.eere.energy.gov/water/hydro_history.html.

Water Power Program: Types of Hydropower Plants. (n.d.). U.S. Department Of Energy, Energy Efficiency and Renewable Energy. Retrieved on May 29, 2012, from http://www1.eere.energy.gov/water/hydro_plant_types.html.

Watson, B. (2012). "How the (Finally Ended) Corn Ethanol Subsidy Made Us Fatter." *Daily Finance*. January 4. Retrieved on January 20, 2012, from http:// www.dailyfinance.com/2012/01/04/how-the-finally-ended-corn-ethanol- subsidy-made-us-fatter/.

Webster, L. (2009). "Diesel Cars in Europe vs. America—Why Diesel Vehicles Are Expensive in US." *Popular Mechanics*. September 10. Retrieved on December 20, 2011, from http://www.popularmechanics.com/cars/alternative-fuel/ diesel/4330313.

Why Hydro. (n.d.). National Hydropower Association. Retrieved on February 13, 2012, from http://hydro.org/why-hydro/.

Wild Free Roaming Horse & Burro Act. (2010). *America's Wild Horse Preservation Campaign*. Retrieved on January 4, 2012, from http://www.wildhorse- reservation.org/resources/1971_act.html.

Wilderness Act of 1964, The. (n.d.). *The Wilderness Society*. Retrieved on February 8, 2012, from http://wilderness.org/content/wilderness-act-1964.

Wildlife and Fisheries: Endangered and Threatened Wildlife and Plants, 50 CFR 17.3. (2002).

Williams, C.F., Reed, M.J., Mariner, R.H., DeAngelo, J., and Galanis, S.P. Jr. (2008). Assessment of moderate- and high-temperature geothermal resources of the United States. U.S. Geological Survey fact sheet 2008–3082, 4. Washington, DC: U.S. Department of the Interior.

Wind Energy and the Environment—Energy Explained, Your Guide To Under-standing Energy. (2011). U.S. Energy Information Administration. July 29. Retrieved on February 11, 2012, from http://www.eia.gov/energyexplained/ index.cfm?page=wind_environment.

Wind Energy Basics. (n.d.). Wind Energy EIS Public Information Center. Retrieved on February 11, 2012, from http://windeis.anl.gov/guide/basics/index.cfm.

Wind Energy Explained. (2011). U.S. Energy Information Administration. May 20. Retrieved on January 28, 2012, from www.eia.gov/energyexplained/index. cfm?page=wind_home.

Wind—Energy explained, Your Guide to Understanding Energy. (2012). U.S. Energy Information Administration. March 15. Retrieved on May 29, 2012, from www.eia.gov/energyexplained/index.cfm?page=wind_home.

Wind Energy Benefits, Wind Powering America Fact Sheet Series. (2005). Department of Energy. April. Retrieved on February 3, 2012, from http://www.nrel. gov/docs/fy05osti/37602.pdf.

Wind in power: 2009 Statistics (2010). European Wind Energy Association. February. Retrieved on February 8, 2012, from http://ewea.org/fileadmin/ewea_documents/documents/statistics/100401_General_Stats_2009.pdf.

Winds of Change. (2010) Dir. Peter Wiesner. Institute of Electrical and Electronic Engineers Production, 2011. Film.

Wind Power: Capacity Factor, Intermittency, and what happens when the wind doesn't blow?. Wind Power on the Community Scale. (2011). Renewable Energy Research Laboratory, University of Massachusetts at Amherst. Retrieved on January 5, 2012, from http://www.umass.edu/windenergy/publications/published/communityWindFactSheets/RERL_Fact_Sheet_2a_Capacity_Factor.pdf.

Woody, T. (2009). "Judge Halts Windfarm Over Bats." *New York Times*. December 10. Retrieved on November 24, 2011, from http://green.blogs.nytimes. com/2009/12/10/judge-halts-wind-farm-over-bats/.

Young, T. (2011). "Malaysian Palm Oil Destroying Forests, Report Warns." *Guardian*. February 2. Retrieved on February 29, 2012, from http://www.guardian. co.uk/environment/2011/feb/02/malaysian-palm-oil-forests.

Zellmer, S. (2011). "Mudslinging on the Missouri: Can Endangered Species Survive the Clean Water Act." *Journal of Agricultural Law* 16: 88–115.

Zoellick, R. (2008). "World Bank Chief: Biofuels Boosting Food Prices." *National Public*. April 11. Retrieved on December 19, 2011, from http://www.npr.org/templates/story/story.php?storyId=89545855.

Zwaan, B., and Rabl, A. (2003). "Prospects for PV: a learning curve analysis." *Solar Energy* 74 (1): 19–31. Retrieved on January 26, 2012, from http://web. me.com/arirabl/Site/Publications_files/vdZwaan+Rabl03%20ProspectsPV-SE. pdf.

Index